Advances in Industrial Control

More information about this series at http://www.springer.com/series/1412

Yihui Wang · Bin Ning · Ton van den Boom
Bart De Schutter

Optimal Trajectory Planning and Train Scheduling for Urban Rail Transit Systems

Springer

Yihui Wang
State Key Laboratory of Rail Traffic Control
and Safety
Beijing Jiaotong University
Beijing
China

Bin Ning
State Key Laboratory of Rail Traffic Control
and Safety
Beijing Jiaotong University
Beijing
China

Ton van den Boom
Delft Center for Systems and Control
Delft University of Technology
Delft
The Netherlands

Bart De Schutter
Delft Center for Systems and Control
Delft University of Technology
Delft
The Netherlands

MATLAB® is a registered trademark of The MathWorks, Inc., 3 Apple Hill Drive, Natick, MA 01760-2098, USA, http://www.mathworks.com.

ISSN 1430-9491 ISSN 2193-1577 (electronic)
Advances in Industrial Control
ISBN 978-3-319-30888-3 ISBN 978-3-319-30889-0 (eBook)
DOI 10.1007/978-3-319-30889-0

Library of Congress Control Number: 2016933235

Printed on acid-free paper

This Springer imprint is published by Springer Nature
The registered company is Springer International Publishing AG Switzerland

Series Editors' Foreword

The series *Advances in Industrial Control* aims to report and encourage technology transfer in control engineering. The rapid development of control technology has an impact on all areas of the control discipline. New theory, new controllers, actuators, sensors, new industrial processes, computer methods, new applications, new philosophies…, new challenges. Much of this development work resides in industrial reports, feasibility study papers, and the reports of advanced collaborative projects. The series offers an opportunity for researchers to present an extended exposition of such new work in all aspects of industrial control for wider and rapid dissemination.

The transportation sector of industry is the focus of much investment in infrastructure, vehicles, and technology. It is a sector highly dependent on control engineering for its operation, safety, reliability, and reduced environmental pollution. The *Advances in Industrial Control* monograph series tries to follow developments in this field by seeking texts that show the interaction between control (theory and applications) and the latest developments in transportation technology. It is interesting to examine the trends in various branches of transportation technology and view how the recent contributions to the *Advances in Industrial Control* monograph series are reporting such developments.

Aerospace Control Systems

Aerospace systems cannot exist without control and consequently there are always new developments in this field. There are two trends in terrestrial aerospace control systems that find expression in the series. The first is new aerospace structures as exemplified by the monographs:

- *Active Control of Flexible Structures* by Alberto Cavallo, Giuseppe De Maria, Ciro Natale, Salvatore Pirozzi (ISBN 978-1-84996-280-3, 2010); and
- *Modeling and Control for a Blended Wing Body Aircraft* edited by Martin Kozek and Alexander Schirrer (ISBN 978-3-319-10791-2, 2015).

The second trend is that of control for rotorcraft aerial vehicles and in particular, for autonomous rotorcraft vehicles. Recent entries into the series include:

- *Unmanned Rotorcraft Systems* by Guowei Cai, Ben M. Chen, Tong Heng Lee (ISBN 978-0-85729-634-4, 2011);
- *Quad Rotorcraft Control* by Luis R. García Carrillo, Alejandro E. Dzul López, Rogelio Lozano, and Claude Pégard (ISBN 978-1-4471-4398-7, 2013); and
- *Smart Helicopter Rotors* by Ranjan Ganguli, Dipali Thakkar and Sathyamangalam R. Viswamurthy (ISBN 978-3-319-24766-3, 2015).

Marine Control Systems

As one of the oldest forms of transportation, control systems in the marine field are mature. However, even these systems are not immune from new concepts in control theory as the following monographs demonstrate:

- *Robust Control of Diesel Ship Propulsion* by Nikolaos Xiros (ISBN 978-1-85233-543-4, 2002);
- *Ship Motion Control* by Tristan Perez (ISBN 978-1-85233-959-3, 2005); and
- *Control of Ships and Underwater Vehicles* by Khac Duc Do and Jie Pan (ISBN 978-1-84882-729-5, 2009).

Traffic and Automobile Control Systems

For some years now the American Control Conference has fielded special sessions on automobile control systems. These sessions reflect the technological change caused by having new sensors, computers, and concomitant software as integral components of the automobile system. The series recently fielded three monographs in this growing "smart" car technology:

- *Dry Clutch Control for Automotive Applications* by Pietro J. Dolcini, Carlos Canudas de Wit and Hubert Béchart (ISBN 978-1-84996-067-0, 2010);
- *Active Braking Control Systems Design for Vehicles* by Sergio M. Savaresi and Mara Tanelli (ISBN 978-1-84996-349-7, 2010); and
- *Optimal Control of Hybrid Vehicles* by Bram de Jager, Thijs van Keulen and John Kessels (ISBN 978-1-4471-5075-6, 2015).

The ultimate outcome of "smart" car control technology must surely be the driverless car.

However, the proliferation of petrol- and diesel-fuelled automobiles and the rise of the long-distance commuter have created the problem of very severe air pollution in many global conurbations, for example in, Beijing, Delhi, Paris, Taipei, and London. Short-term solutions have involved the use of traffic-management schemes to reduce the number of vehicles entering these areas. The series was fortunate to publish a seminal monograph in this traffic-management field in 1998, namely:

- *Feedback Control Theory for Dynamic Traffic Assignment* by Pushkin Kachroo and Kaan M.A. Özbay (ISBN 978-1-4471-1209-9, 1998).

Bus Urban Transit Systems

The long-term solution being pursued in many global locations involves investment, and the construction and optimization of modern, safe, and reliable urban transit systems. These systems can involve a number of different transport modalities such as trams, buses, and overground and underground Metro systems. There is potential for control engineers to play a significant role in the development of such systems. An important monograph in the series that used recent control theory to analyze urban bus transportation was *Hybrid Predictive Control for Dynamic Transport Problems* by Alfredo A. Núñez, Doris A. Sáez and Cristián E. Cortés (ISBN 978-1-4471-4350-5, 2012).

Rail and Rail-Based Urban Transit Systems

Long-distance rail travel and rail-based urban transit systems are recognized as key solutions to moving large numbers of people from place to place with high levels of safety and reliability, and with minimized environmental pollution. Many new developments and upgrades of old systems are active worldwide. In 1995, the series was fortunate to publish the seminal monograph for rail control systems, *Energy-Efficient Train Control* by Philip G. Howlett and Peter J. Pudney (ISBN 978-1-4471-3086-4, 1995). Currently, these authors are involved with Amie R. Albrecht in a second edition of this monograph. In this new version the authors also recognize the importance of treating rail-based urban transit systems.

The survey ends with this new monograph addition to the series: *Optimal Trajectory Planning and Train Scheduling for Urban Rail Transit Systems* by Yihui Wang, Bin Ning, Ton van den Boom, and Bart De Schutter. This monograph is sharply focused on the optimization problems of urban rail transit systems and will be of interest to a wide range of readers from the control and transportation communities. It describes the challenge that exists for control engineers and, we hope it leads the way to more contributions in this important technological area. It is an excellent entry into the *Advances in Industrial Control* monograph series.

Industrial Control Centre
Glasgow
Scotland, UK

M.J. Grimble
M.A. Johnson

Preface

Energy efficiency in transportation systems is becoming more and more important because of environmental concerns and for reasons of cost. For the Beijing urban rail transit system, the power consumption in 2008 was 6.5 million kWh, which was 1 % of the total power consumption of the city. It is predicted that the power consumption of Beijing urban rail transits system will rise to 13.9 million kWh in 2015, which would then account for 1.2 % of the total power consumption. Furthermore, the energy consumption for the operation of trains is about 40–50 % of the total power consumption for urban rail transit systems. Therefore, the reduction in energy consumption for the operation of trains is one of the key objectives in the operation of railway systems. Meanwhile, the interest of railway operators in energy efficiency has been increasing more and more in recent years, as even a small improvement in the energy consumption can make railway operators save a lot of money.

Furthermore, with the increase in passenger demand in large cities like Beijing, Shanghai, Tokyo, New York, and Paris, urban rail transit systems play an increasing role in the efficiency and sustainability of the overall transportation system. Nowadays the operation of trains in urban rail transit systems is characterized by a high frequency, where the minimum headway between two successive trains is usually 2–5 min, which could be reduced to even 90 s with the development of advanced train control systems. When trains are operated at such a high frequency, their scheduling based on the passenger demand becomes more and more important for passenger satisfaction and for the reduction of operation costs. Passenger satisfaction depends on the waiting times, in-vehicle times, and the number of transfers, while the operation costs are determined by the number of train services and the energy consumption of train operations.

The aim of the book is to determine and develop mathematical models and solution approaches to shorten the travel time of passengers and to reduce energy consumption in urban rail transit systems. The content of Chaps. 1–7 can be summarized as follows:

Chapter 1: A brief introduction to railway operations and the structure of this book are presented.

Chapter 2: The background material and literature review on the operation of trains and on urban rail train scheduling are presented.

Chapter 3: The optimal trajectory planning problem for the operation of a single train is considered. The varying line resistance, variable speed restrictions, and varying maximum traction force are included in the problem definition. The objective function is a trade-off between the energy consumption and the riding comfort. Two approaches are proposed to solve this optimal control problem:

- A pseudospectral method, i.e., a state-of-the-art method for optimal control, is applied for the first time in optimal train control, where the optimal trajectory planning problem is recast into a multiple-phase optimal control problem.
- The optimal trajectory planning problem is reformulated as a mixed integer linear programming (MILP) problem by approximating the nonlinear terms in the problem by piecewise affine functions. The resulting MILP problem can be solved efficiently by existing commercial and open source solvers.

The main conclusion of the chapter is that if the available time for computing the solution is large enough, the pseudospectral method is recommended since it has a higher performance. Otherwise, the MILP approach can be applied to obtain the optimal train trajectory.

Chapter 4: We investigate the optimal trajectory planning problem for multiple trains under fixed block signaling systems and moving block signaling systems. The constraints caused by the leading train in a fixed or moving block signaling system are included into the model for the operation of trains. Two solution approaches are proposed to solve the resulting optimal control problem for multiple trains:

- The greedy approach optimizes the trajectory of the leading train first and then based on the optimal trajectory of the leading train, the trajectory planning problem for the following train is solved. The MILP approach and the pseudospectral method are employed to solve the optimal trajectory for the leading train and the following train.
- The simultaneous approach optimizes the trajectory of all the trains in the problem formulation at the same time, where the MILP approach and the pseudospectral method are also applied.

In the simulation experiment, we show that the simultaneous MILP/pseudospectral approach yields a better performance but requires a higher computation time compared to the greedy MILP/pseudospectral approach.

Chapter 5: The real-time train scheduling problem for an urban rail transit line is investigated with the aim of minimizing the total travel time of passengers and the energy consumption of the operation of trains. The train scheduling problem in this chapter is based on origin-destination-independent (OD-independent) passenger demands, which means that the passenger arrival rates at stations do not depend on the origins and destinations of passengers but only on the number of passengers that arrived. With consideration of OD-independent passenger demands, the optimal departure times, running times, and dwell times are obtained by solving the scheduling problem for the urban rail transit system. Several approaches are proposed to solve this problem:

- A new iterative convex programming (ICP) approach is proposed to solve the train scheduling problem.
- Nonlinear programming approaches (sequential quadratic programing (SQP) and pattern search in particular), a mixed integer nonlinear programming approach, and an MILP approach are also applied to solve the problem.

We find that the ICP approach provides the best trade-off between performance and computational complexity comparing with other alternative solution approaches.

Chapter 6: We consider the train scheduling taking constant origin-destination-dependent (OD-dependent) passenger demands into account for an urban rail transit line, where the origins and destinations of passengers are considered in the passenger arrival rates at stations. A stop-skipping strategy is adopted to reduce the passenger travel time and the energy consumption. The train scheduling problem results in a mixed integer nonlinear programming problem, where two solution approaches are proposed:

- A bi-level approach is considered to solve the train scheduling problem, where the higher level optimizes the binary variables and the lower level solves a nonlinear non-convex problem for each combination of binary variables.
- We also propose a limited bi-level approach that first applies a threshold method to obtain a good initial solution for the problem and then limits the search space of the variables to speed up the optimization process.

Experiment indicates that the bi-level approach has better control performance, but the limited bi-level approach can provide an acceptable solution with much less computation time.

Chapter 7: An event-driven model is proposed for the train scheduling of an urban rail transit network, where a time-varying OD-dependent passenger demand, splitting of passenger flows, and passenger transfer behavior at transfer stations are included. This event-driven model involves three types of events, i.e., departure events, arrival events, and passenger arrival rate change events (which are used to describe the changes of passenger arrival rates at platforms due to time-varying passenger demand, passenger transfers, etc.) at platforms. The train scheduling problem that is constructed based on the event-driven model is a real-valued nonlinear non-convex programming problem. Several solution approaches, such as sequential quadratic programming (SQP), pattern search method, mixed integer linear programming, and genetic algorithms, can be applied to this train scheduling problem.

The simulation results of the case study show that the SQP method provides a better trade-off between control performance and computational complexity than the genetic algorithm.

The models and approaches proposed in this book can be applied in urban rail transit systems to improve the train scheduling and train control in practice. In addition, they can also be adapted and extended to control and to optimize other transportation systems, such as bus transit systems, multi-car elevator systems, and automated guided vehicles.

This work is supported by the State Key Laboratory of Rail Traffic Control and Safety, Beijing Jiaotong University and the Delft Center for Systems and Control, Delft University of Technology. The authors thank Prof. Tao Tang for his support in publishing this book.

Beijing (China)
December 2015

Yihui Wang
Bin Ning
Ton van den Boom
Bart De Schutter

Contents

Notation

Below is a list of the most frequently used symbols and acronyms that occur in this book.

Symbols

Chapter 3

m	Mass (kg) of a train
g	Gravitational acceleration (m/s^2)
ρ	Rotating mass factor (–)
v	Speed (m/s) of a train
$V_{\max}(s)$	Speed limit (m/s) along the track
s	Position (m) of a train
u	Traction or braking force (N) working on a train
$u_{\max}^{\text{acc}}, u_{\max}^{\text{dec}}$	Maximum traction and braking force (N)
$R_b(v)$	Roll resistance and air resistance (N) of a train
$R_l(s, v)$	Line resistance (N) caused by track grade, curves, and tunnels
$\alpha(s)$	Slope along the track (rad)
$r(s)$	Radius (m) of the curve along the track
$l_t(s)$	Length (m) of tunnels along the track
$f_c(r(s))$	Curve resistance (N)
f_t	Tunnel resistance (N)
T	Given running time (s) for a train
s_k	Boundary (m) of discrete space interval k
Δs_k	Length (m) of discrete space interval k
$E(k)$	Kinetic energy (J) of a train at position s_k
$E_{\min}$	Minimum kinetic energy (J)
$t(k)$	Passing time (s) of a train at position s_k

Chapter 4

$v_{\max}$	Speed code (m/s) for a green signal aspect in a three-aspect FBS system
v_{yellow}	Speed code (m/s) for a yellow signal aspect in a three-aspect FBS system
$v_{\min}$	Speed code (m/s) for a red signal aspect in a three-aspect FBS system
$H_{\mathrm{min,FBS}}$	Minimum headway (s) between trains in an FBS system
$H_{\mathrm{min\ MBS}}$	Minimum headway (s) between trains in an MBS system
$s_{\mathrm{FB},m}$	Boundary (m) of fixed block section *m*
L_{a}	Length (m) of a block in an FBS system
$L_{\mathrm{t}}^{\mathrm{L}}$	Length (m) of the leading train
L_{s}	Length (m) of the secure section
S_{SM}	Safety margin (m) for braking of the following train
$v_{\max}^{\mathrm{F}}$	Maximum speed (m/s) of the following train
$t_{\mathrm{r}}^{\mathrm{F}}$	Reaction time (s) of the driver and equipment of the following train
$L_{\mathrm{r}}^{\mathrm{F}}$	Distance (m) that the following train may travel during the reaction time
$a_{\mathrm{dec}}^{\mathrm{F}}$	Deceleration (m/s^2) of the following train
$t_{\mathrm{dec}}^{\mathrm{F}}$	Deceleration time (s) of the following train
$t_{\mathrm{d}}^{\mathrm{L}}$	Station dwell time (s) of the leading train
$v^{\mathrm{F}}(t)$	Speed (m/s) of the following train at time instant *t*
$a_{\mathrm{acc}}^{\mathrm{L}}$	Acceleration (m/s^2) of the leading train
$t_{\mathrm{safe}}^{\mathrm{F}}$	Safety time margin (s) caused by safety distance margin and train length

Chapter 5

i	Train number
j	Station number
s_j	Track section length (m) between station *j* and station *j* + 1
$d_{i,j}$	Departure time (s) of train *i* at station *j*
$a_{i,j}$	Arrival time (s) of train *i* at station *j*
$\tau_{i,j}$	Dwell time (s) of train *i* at station *j*
$r_{i,j}$	Running time (s) of train *i* at station *j*
$v_{i,j}$	Train speed (m/s) of the holding phase for train *i* on segment *j*
$a_{i,j}^{\mathrm{acc}}$	Acceleration (m/s^2) for train *i* on segment *j*
$a_{i,j}^{\mathrm{dec}}$	Deceleration (m/s^2) for train *i* on segment *j*
h_0	Minimum headway between two successive trains
k_{1i}, k_{2i}, k_{3i}	Resistance coefficients for train *i*
λ_j	Passenger arrival rate (passenger/s) at station *j*
ρ_j	Passenger alighting proportion when a train arrives at station *j*

$C_{i,\max}$	Maximum capacity (passengers) of train i
$w_{i,j}^{\text{wait}}$	Number of passengers waiting for train i at station j
$w_{i,j}$	Number of passengers left immediately after the departure of train i at station j
$n_{i,j}^{\text{remain}}$	Remaining capacity (passengers) of train i at station j after the passenger alighting process
$n_{i,j}^{\text{board}}$	Number of passengers boarding train i at station j
$n_{i,j}^{\text{alight}}$	Number of passengers alighting from train i at station j
$n_{i,j}$	Number of passengers on train i when it departs from station j
$\tau_{i,j,\max}$, $\tau_{i,j,\min}$	Maximum and minimum dwell time (s)
$\alpha_{1,d}, \ldots, \alpha_{4,d}$	Coefficients of the minimum dwell time
n_{door}	Number of doors of the train
$t_{\text{wait},i,j}$	Passenger waiting time (s) at station j for train i
$t_{\text{in-vehicle},i,j}$	Passenger in-vehicle time (s) at station j for train i
$E_{i,j}^{\text{acc}}$	Energy consumption (J) of the acceleration phase at station j for train i
$E_{i,j}^{\text{hold}}$	Energy consumption (J) of the holding phase at station j for train i
$E_{i,j}^{\text{dec}}$	Energy consumption (J) of the deceleration phase at station j for train i
$t_{\text{total,nom}}$	Nominal value (s) of the travel time of passengers
$E_{\text{total,nom}}$	Nominal value (J) of the total energy consumption

Chapter 6

S_{skip}	Skipping set
$h_{0,\text{dep}}$	Minimum departure headway (s) at terminal station
$h_{0,\text{arr}}$	Minimum arrival headway (s) at terminal station
$\tau_{0,\min}$	Minimum dwell time (s) at terminal station
C_0^{ter}	Capacity (passengers) of the terminal station
$y_{i,j}$	Binary variable to indicate whether train i stops at station j or not
$\lambda_{j,m}$	Passenger arrival rate (passengers/s) at station j for passenger with destination m
$w_{i,j,m}^{\text{wait}}$	Number of passengers with destination m waiting for train i at station j
$w_{i,j,m}$	Number of passengers with destination m remaining at station j immediately after the departure of train i
$w_{i,j}^{\text{want-to-board}}$	Number of passengers who want to board train i at station j
$n_{i,j,m}^{\text{board}}$	Number of passengers with destination m boarding train i at station j

Chapter 7

S_{ln}	Set of urban rail transit lines
S_{sta}	Set of stations
S_{pal}	Set of platforms
p	Platform index
$\mathfrak{p}^{\mathrm{pla}}(p)$	Predecessor of platform p
$\mathfrak{s}^{\mathrm{pla}}(p)$	Successor of platform p
$\mathfrak{p}^{\mathrm{tra}}(i)$	Predecessor of train i
$\mathfrak{s}^{\mathrm{tra}}(i)$	Successor of train i
e_k	Event k
t_k	Time instant (s) at which event e_k occurs
$Y_{\mathrm{type},k}$	Event type (departure, arrival, λ-change) of event k
i_k	Index of the train corresponding to event e_k
p_k	Index of the platform corresponding to event e_k
$\lambda_{j,m}^{\mathrm{station}}$	Passenger arrival rates (passengers/s) at station j with destination m
$\beta_{p,m}^{\mathrm{station}}$	Splitting rate of passenger flows that arrive at station j with destination m and that go to platform p
$a_{0,p}^{\mathrm{walk}}, b_{0,p}^{\mathrm{walk}}$	Coefficients for average walking time from the entrance to platform p
$\theta_p^{\mathrm{walk-in}}$	Average walking time (s) for passengers from entrances to platform p
$t_p^{\mathrm{walk-in}}$	Total walking time (s) for passengers from entrances to platform p
$\beta_{i,p,m}^{\mathrm{train}}$	Splitting rate to platform p for passengers that are on board of train i and have destination m
$t_p^{\mathrm{walk-out}}$	Total walking time (s) for passengers from platform p to exit station
$a_{p,p'}^{\mathrm{walk}}, b_{p,p'}^{\mathrm{walk}}$	Coefficients for average walking time from platform p to platform p'
$\theta_{i,p,p'}^{\mathrm{walk}}$	Average walking time (s) for transfer passengers alighting from train i from platform p to platform p'
$a_{p,p'}^{\mathrm{duration}}, b_{p,p'}^{\mathrm{duration}}$	Coefficients for the duration time of the transfer process from platform p to platform p'
$\theta_{i,p,p'}^{\mathrm{duration}}$	Duration time (s) for the transfer process from platform p to platform p' for train i
$w_{p_k}^{\mathrm{wait,before}}$	Number of passengers waiting at platform p_k before event e_k occurs
$w_{p_k}^{\mathrm{wait,after}}$	Number of passengers waiting at platform p_k immediately after event e_k

Abbreviations

The following abbreviations are used in this book:

ATO	Automatic train operation
ATP	Automatic train protection
ATS	Automatic train supervision
DP	Dynamic programming
FBS	Fixed blocking signaling
ICP	Iterative convex programming
MBS	Moving blocking signaling
MILP	Mixed integer linear programming
MINLP	Mixed integer nonlinear programming
PWA	Piecewise-affine
SQP	Sequential quadratic programming
OD-dependent	Origin-destination-dependent
OD-independent	Origin-destination-independent

Chapter 1
Introduction

Abstract Urban rail traffic plays a key role in public transportation since it combines high transport capacity and high efficiency. More specifically, a safe, fast, punctual, energy-efficient, and comfortable railway system is important for the economic, environmental, and social objectives of a country or a city. The main focus of this book is on saving energy in railway operations and on enhancing the passenger satisfaction, which can be achieved via optimal trajectory planning for trains and passenger demand-oriented train scheduling by utilizing several mathematical tools and programming methods like MILP, SQP, etc. In this chapter, we provide a brief introduction to railway operations and then present the structure of this book.

1.1 A Brief Introduction on Railway Operations

A railway system consists of three essential elements: infrastructure (like tracks, stations, signaling equipment, etc.), rolling stock with locomotives and cars or electric multiple units (EMUs), and customers (like passenger and/or goods demands) [1–3]. The design, construction, and operation of the infrastructure and rolling stock are affected by the customers. Passenger railway systems could be classified into interurban railway systems (or standard railway systems) and urban rail transit systems (such as metros and subways). Rail infrastructure is a limited resource in interurban rail transit systems, where lines overlap or cross with each other and trains usually overtake or meet each other. On the other hand, in urban rail transit systems, the lines are separated from each other and each direction of the line has a dedicated infrastructure. Moreover, in principle trains do not overtake and meet each other in urban rail transit systems.

The optimal trajectory planning (i.e., speed profile calculation) methods for the operation of trains proposed in this book can be applied both for interurban railway systems and urban rail transit systems. However, train scheduling approaches we present here are focused on urban rail transit systems.

In railway systems, the operation of trains is in general controlled through a hierarchical control framework with five levels, i.e., scheduling, real-time (re)scheduling, remote traffic control, interlocking and signaling, and train and infrastructure control

Y. Wang et al., *Optimal Trajectory Planning and Train Scheduling for Urban Rail Transit Systems*, Advances in Industrial Control,
DOI 10.1007/978-3-319-30889-0_1

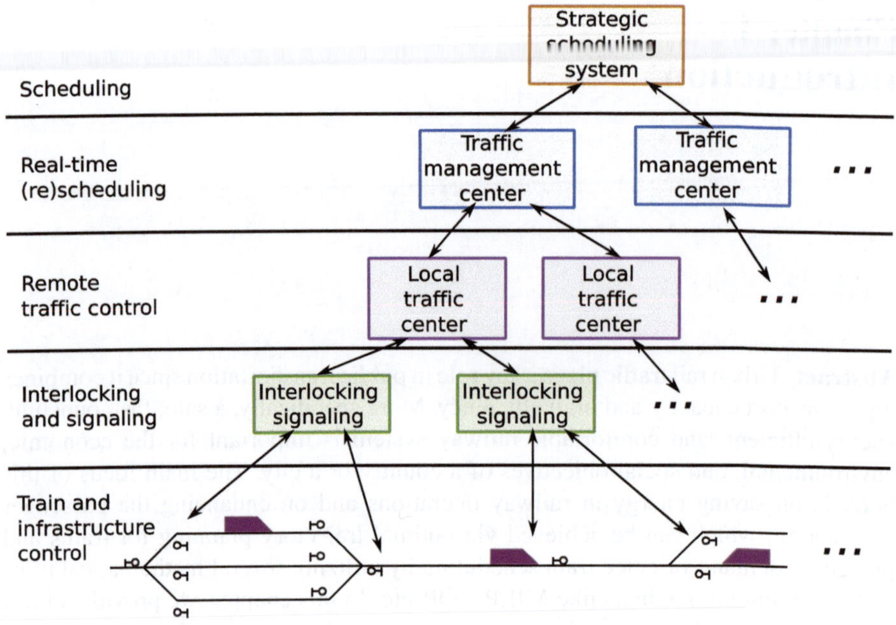

Fig. 1.1 Hierarchical structure of the railway operations (Adopted from [4])

(see Fig. 1.1) [4]. The scheduling process for railway systems involves a complex procedure that includes demand analysis, line planning, train scheduling, rolling stock planning, and crew scheduling (see Sect. 2.3 for more detailed information). When delays, interruptions, or failures, etc. occur, dispatchers in the traffic management centers supervise the railway network and they resolve the conflicts through rescheduling. The decisions of the dispatchers are then forwarded to the remote traffic control level, where the local traffic centers set the routes and issue the speed limits for trains through interlocking systems and signaling systems. Moreover, there also exists an opposite information flow: the information of train operations is gathered automatically at the train control level and forwarded upward to the traffic management systems.

Safety is an important issue for railway systems, where signaling systems and interlocking systems are employed to prevent collisions between trains and ensure safe train movements. There are two principles for signaling systems, viz. fixed block signaling principle and moving block signaling principle (for more information see Sect. 2.1.2). In practice, advanced signaling systems for train control and safety have been developed, such as the European train control system for interurban railway systems [1] and communication-based train control systems for urban rail transit systems [5]. More specifically, the European train control system has three levels, where level 1 and 2 are based on the fixed block signaling principle and level 3 is based on the moving block signaling principle. For modern urban rail transit systems, the moving block signaling principle is often implemented in the communication-based train control system and the fixed block signaling system is also implemented as a back-up. The architecture of the communication-based train control system may involve automatic train protection (ATP), automatic train operation (ATO), and automatic train supervision (ATS). In particular, the ATP system is used to guarantee the safety of the operation of trains, where deviating driving behaviors (like over-speeding or passing red signals) caused by drivers or ATO systems would trigger an emergency brake. Furthermore, the ATO system can control the traction and braking effort automatically to operate trains between stations (see Sect. 2.1.1 for more detailed information). In addition, there are also some driver assistance systems that have been developed in practice to assist drivers to drive the train optimally, such as FreightMiser [6], Metromiser [6], and driving style manager [7].

Train scheduling is crucial for railway operations since it has a significant impact on infrastructure usage and punctuality. For interurban railway systems, the timetables are usually published to provide trip information to passengers. The drawback of fixed timetables is that adjustments (due to e.g., changes of demand) are difficult to implement. However, in urban rail transit systems, trains are operated with high frequencies and rail transport operators often do not publish the train schedule to passengers but only provide some real-time information, such as that a train will arrive within 2 min. Furthermore, the passenger demands for urban rail transit systems may change rapidly with the time of the day or due to some special events. Hence, rail transit operators could schedule trains in real time based on the passenger demands to increase passenger satisfaction with short and reliable travel times.

The aim of the book is to determine and develop mathematical models and solution approaches to shorten the travel time of passengers and to reduce energy consumption in railway systems. With respect to the five-level hierarchical control framework discussed above (cf. Fig. 1.1), scheduling, real-time (re)scheduling, and train control are closely related to the energy consumption of trains and the travel time of passengers, while the remote traffic control and the signaling and interlocking systems are responsible for the safety of the operation of trains. In the train control level, trains are operated by drivers or ATO systems under the supervision of ATP systems. In addition, trains are assumed to run according to the schedule given by the train scheduling or real-time (re)scheduling, i.e., the fixed running times for trains between two consecutive stations and fixed dwell time at stations. Generally, the scheduling of trains means to generate an off-line timetable for the railway system. Based on the existing timetable data, real-time rescheduling is used to handle route conflicts due to train delays or incidents. In this book, real-time scheduling means that there is no existing timetable or constant headways, but the schedule of trains is optimized in a rolling horizon way taking passenger demands and operation costs into consideration. The travel time of passengers is considered in the train scheduling, where the running times and dwell times of trains are determined. Note that the energy consumption depends on the running times of trains, so the energy saving can also be dealt with in the train scheduling layer. In this book, the following two topics are investigated:

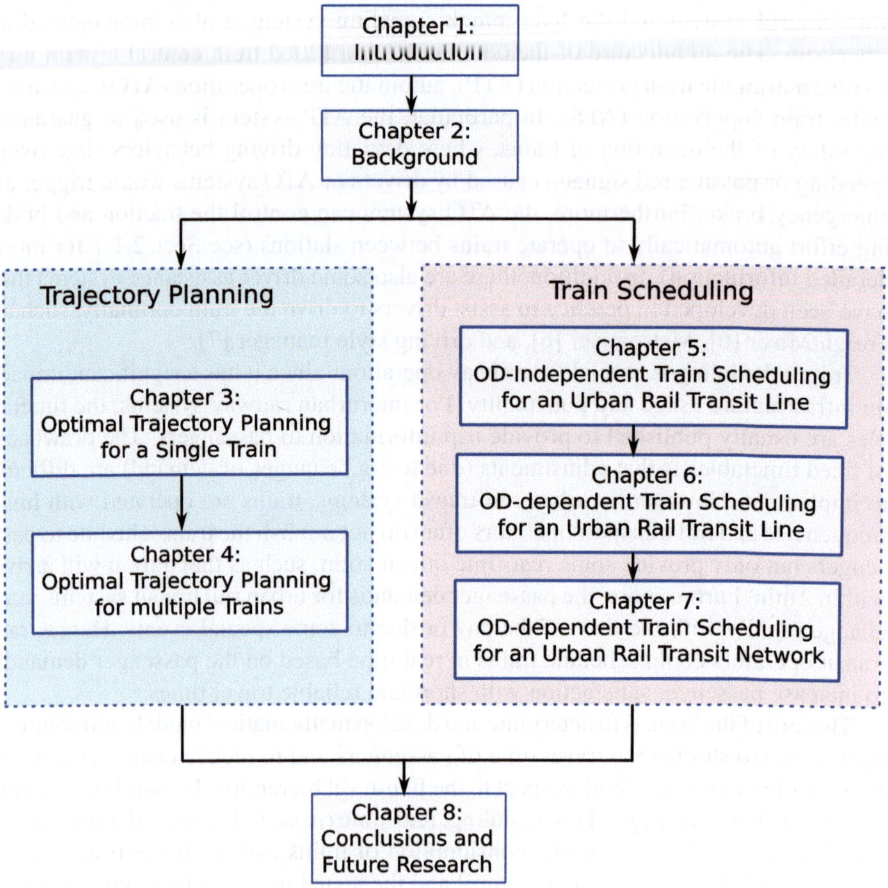

Fig. 1.2 The road map of this book

- Trajectory planning. A nonlinear model for the operation of trains is derived and several approaches are presented to calculate the optimal trajectories in an energy-efficient way for trains based on a given train schedule.
- Train scheduling. To shorten the travel time of passengers and to reduce the energy consumption, we develop a train scheduling model for urban rail transit systems and optimization approaches to find a balanced trade-off between total passenger travel time and the operation cost of the rail transport operator.

1.2 Book Outline

A road map of the book is presented in Fig. 1.2, which clarifies the connections between the chapters. According to the road map, readers interested in optimal trajectory planning could read the book in the following order: This chapter; Sects. 2.1,

2.1.2, and 2.4 of Chap. 2; Chaps. 3, 4, and 8. Those interested in the train scheduling problem for urban rail transit system could read the book in the following order: This chapter; Sects. 2.3 and 2.4 of Chap. 2; Chaps. 5, 6, 7, and 8.

This chapter gives the motivation and a general introduction to the topic of the book. Chapter 2 presents the background of the operation of trains and the train scheduling process and summarizes the related research work in the literature. Chapters 3 and 4 investigate on optimal trajectory planning for a single train and for multiple trains, respectively. Chapters 5 and 6 focus on train scheduling for an urban rail transit line, where the passenger characteristics are described in different ways. The train scheduling problem for urban rail transit networks is formulated in Chap. 7, where the passenger transfers are included. Chapter 8 concludes the book with the main contributions and directions for future research.

References

1. Pachl J (2009) Railway operation and control, 2nd edn. Gorham Printing, Centralia
2. Caimi G (2009) Algorithmic decision support for train scheduling in a large and highly utilised railway network. Ph.D. thesis, ETH Zürich, Switzerland
3. Liebchen C (2006) Periodic timetable optimization in public transport. Ph.D. thesis, Technique University of Berlin, Berlin, Germany
4. Lüthi M (2009) Improving the efficiency of heavily used railway networks through integrated real-time rescheduling. Ph.D. thesis, ETH Zürich, Switzerland
5. IEEE 1474.1 (2004) IEEE Standard for Communications-Based Train Control (CBTC) Performance and Functional Requirements
6. Howlett P, Pudney P (1995) Energy-efficient train control. Advances in industrial control. Springer, London
7. Franke R, Meyer M, Terwiesch P (2002) Optimal control of the driving of trains. Automatisierungstechnik 50:606–614

Chapter 2
Background: Train Operations and Scheduling

Abstract In this chapter, background material and literature review on the operation of trains and on urban rail train scheduling are presented. In Sect. 2.1, the operation of trains is introduced, where the automatic train operation (ATO) system is explained in detail. In addition, a brief introduction to fixed block signaling systems and moving block signaling systems is also given. An overview of optimal control approaches for the trajectory planning of a single train and multiple trains is provided in Sect. 2.2. Section 2.3 introduces the urban rail transit scheduling problem is introduced. This chapter concludes with a short summary in Sect. 2.4.

2.1 Operation of Trains

Nowadays, several dedicated high-speed railway lines and urban rail transit systems with short headways are operated with a high degree of automation [1]. This requires advanced train control systems to fulfill safety and operational requirements, such as the European train control system and communication-based train control systems, which include equipment on board of trains as well as in control centers [2]. Advanced train control systems enable the energy-efficient driving of trains, which becomes more and more important because of the operation costs and environmental concerns [3].

The automatic train operation (ATO) system of an advanced train control system drives the train according to a predefined train trajectory (i.e., a speed profile) [4] to ensure punctuality and energy saving. In addition, signaling systems in train control systems is important for running safety of trains. In this section, we first give a brief introduction to ATO systems and then provide a short introduction to the principle of signaling systems.

2.1.1 Automatic Train Operation

With the development of modern railway systems, automatic train control (ATC) systems have become vital equipment that ensures the running safety, shortens the

Y. Wang et al., *Optimal Trajectory Planning and Train Scheduling for Urban Rail Transit Systems*, Advances in Industrial Control,
DOI 10.1007/978-3-319-30889-0_2

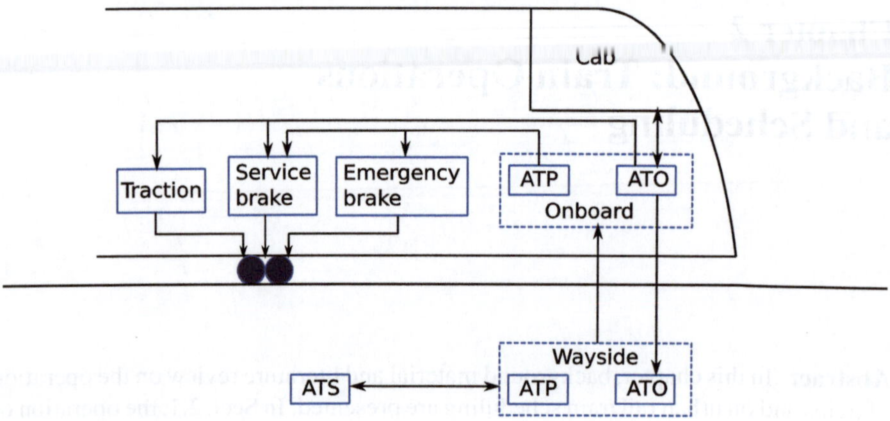

Fig. 2.1 The structure of an advanced automatic train control (ATC) system [5]

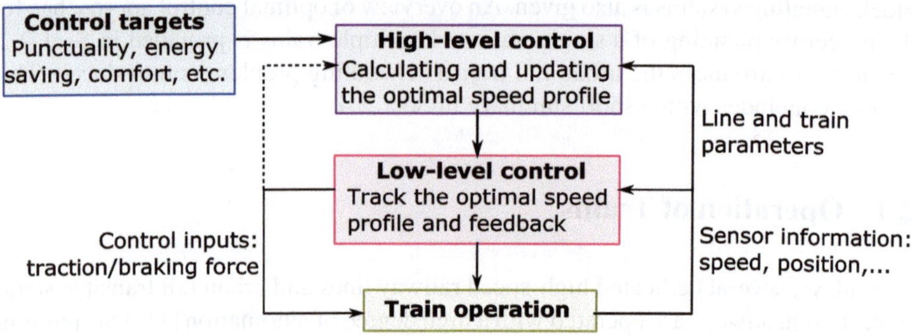

Fig. 2.2 The schematic diagram of the control actions in an ATO system

train headways, and improves the quality of train operations [4]. An advanced automatic train control system could consist of an automatic train protection (ATP) system, an automatic train supervision (ATS) system, and an ATO system as shown in Fig. 2.1 [5]. The onboard ATP system is responsible for supervising the train speed according to the safety speed profile and for applying an appropriate braking force if necessary. In addition, the onboard ATP system also communicates with the wayside ATP system to exchange information (e.g., temporary speed limits and the limits of movement authority (i.e., the maximum position that a train is allowed to move to)) to guarantee the safety of the operations of trains. The ATS system acts as an interface between the operator and the railway system, managing trains according to the specific regulation criteria. The ATO system controls the traction and braking force to keep the train speed under the speed limit established by the ATP system. The ATO system can be used to facilitate the driver or to operate the train in a fully automatic mode; it thus plays a key role in ensuring accurate stopping, operation punctuality, energy saving, and riding comfort [4].

An onboard ATO system consists of two levels of control actions, as conceptually illustrated in Fig. 2.2. The higher level optimizes the optimal speed-position reference trajectory for the operation of the train, where the line resistance, speed limits, maximum traction and braking forces, etc. are taken into account. The low-level control is used to make the train track the preplanned reference trajectory via certain control methods (such as PID control, model predictive control, and robust control). The traction or braking control commands are implemented to the train and information on e.g. the speed and position of the train is collected by the sensors and transferred to the ATO system in real time.

The driving performance including punctuality, energy consumption, etc. strongly depends on the optimal reference trajectory both when the train is partly or fully controlled by the ATO system. In addition, there exist several driver assistance systems to enhance the driving performance of the drivers, such as the FreightMiser, Metromiser, and the driving style manager. The FreightMiser and Metromiser systems [6] were developed by the scheduling and control group of the University of South Australia in order to calculate the optimal reference trajectory and to give advices to the drivers of long-haul trains and suburban trains respectively. That group mainly focused on minimizing the energy consumption through Pontryagin's principle [7]. The driving style manager [8] developed by Bombardier implements dynamic programming to calculate energy-efficient train trajectories, which are then displayed to the train driver. Whenever the train stops at a station, the driving style manager calculates the optimal trajectory to the subsequent station using real-time information. An energy-efficient driving module is integrated in the Trainguard MT communications-based train control (CBTC) system of Siemens, where a simulation-based approach is applied to obtain the energy-optimal trajectories for trains [9]. The interested reader is referred to [10] for more information on the implemented systems.

ATO systems and driver assistance systems are able to take advantage of a pre-computed train speed trajectory. However, if the operational conditions change, the ATO system will calculate an updated optimal trajectory. Therefore, it is important to design efficient algorithms to find the optimal speed-position reference trajectory. In the literature, various algorithms have been developed to optimize the speed trajectory for trains and these algorithms will be reviewed in Sect. 2.2.

2.1.2 *Principles of Signaling Systems*

Block signaling is used to maintain a safe distance between successive trains on the same track. There are two main types of signaling systems, namely fixed block signaling systems and moving block signaling systems. The main principles of those two signaling systems are presented next.

2.1.2.1 Fixed Block Signaling Systems

Fixed block signaling (FBS) systems are commonly used in railway operation systems nowadays [11]. In FBS systems, a track is divided into blocks, the length of which depends on the maximum train speed, the worst-case braking rate, and the

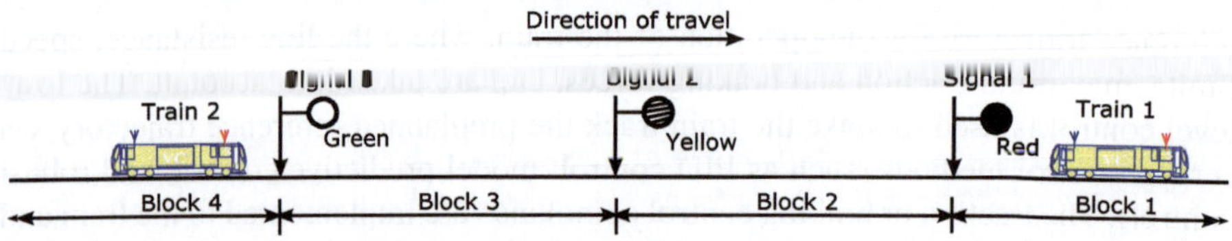

Fig. 2.3 Three aspect fixed block signaling system

number of signal aspects, such as a green, yellow, or red. Each block is exclusively occupied by only one train and the presence of a train within a block is usually detected by the track circuits [12]. Furthermore, blocks are protected by wayside signals (i.e., signals next to the track) or cab signals (i.e., visual signals on board of trains). Wayside signals are still typical in railways, however, cab signals are used more and more, in particular on high-speed lines where wayside signals cannot be watched clearly by drivers because of the high speed. There are one-block signaling and multiple-block signaling in FBS systems [11]. In one-block signaling, the indication of the block signal depends only on the state of the block section after the signal and every block signal must have a distant signal, which is supposed to provide the required approach information. In multiple-block signaling systems, the indication of a block signal depends on the state of two or more subsequent block sections.

A simple example is a two-block signaling system with three aspects, i.e. red, yellow, and green, and which is also called a three-aspect signaling system. Such a three-aspect signaling system on a line equipped with an ATP system is shown as Fig. 2.3. Each block carries an electronic speed code through its track circuit. The speed code data consists of two parts: the authorized-speed code for this block and the target-speed code for the next block. The speed code data is coded by the electronic equipment controlling the track circuitry and is transmitted via tracks. This speed code data is then picked up by antennae on board of the train. If a train tries to enter a zero speed block or an occupied block, or if it enters a section at a speed higher than that authorized by the speed code, the onboard electronics will trigger an emergency brake.

2.1.2.2 Moving Block Signaling Systems

With the increasing operational density in railway systems, railway systems with an FBS system are often suffering from a shortage in transportation capacity. Even though the line capacity of an FBS system can be increased using shorter block lengths, the installation and maintenance cost of the signaling and track equipment may not be justified by the increased capacity. Consequently, moving block signaling (MBS) systems have been proposed to achieve a higher performance.

In an MBS system, the blocks are defined as dynamic safe zones around each train. Regular communication between trains and local traffic centers is needed for knowing the exact locations and speeds of all trains in the area controlled by the local traffic

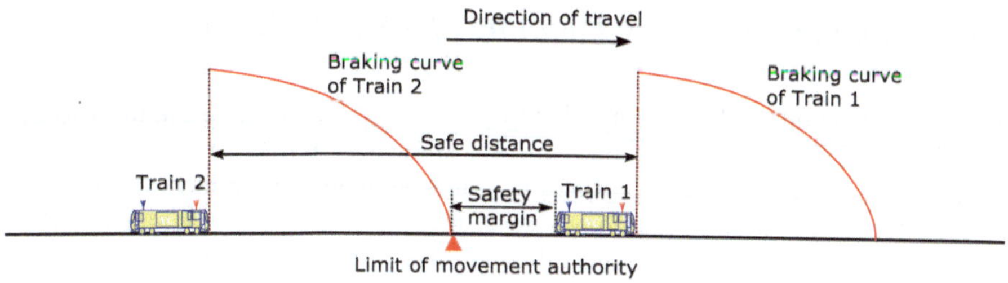

Fig. 2.4 The principle of a pure MBS system

center at any given time. Therefore, compared to an FBS system, an MBS system allows trains to run closer together, thus increasing the transport capacity. The local traffic center computes the so-called limit of movement authority for every train in the area it controls and makes sure that each train will be running at a safe distance with respect to other trains (cf. Fig. 2.4). More specifically, the limit of movement authority represents the maximum position that a train is allowed to move to and it is determined by the tail of the preceding train with a safety margin included. In addition, the limit of movement authority of the following train moves forward continuously as the leading train travels. In the literature, four MBS schemes [13] have been discussed: moving space block signaling, moving time block signaling, pure MBS, and relative MBS. Takeuchi et al. [12] evaluated the first three schemes and compared them with the FBS scheme based on two basic criteria, viz., steady-state performance and perturbed performance. It is concluded that the pure MBS scheme gives the best performance. In addition, Takeuchi et al. [12] stated that the concept of the relative MBS has never been accepted for regular rail traffic even though it is routinely accepted for road traffic. Therefore, we will mainly consider the pure MBS scheme later on in this book. However, the proposed approaches can be extended to other MBS schemes too. Moreover, the pure MBS scheme is the basis of all systems currently implemented in practice [12].

In a pure MBS system, the minimum distance between two successive trains is basically the sum of the instantaneous braking distance required by the following train and a safety margin (which is introduced to avoid collisions even if the leading train comes to a sudden halt) as shown in Fig. 2.4. However, the minimum distance between trains in practice should also take the train length and the running distance during the reaction time of the drivers or ATC systems into account.

2.2 Optimal Trajectory Planning of Trains

In this section, we first give a literature review on the optimal trajectory planning of a single train and then the state-of-the-art on the trajectory planning of multiple trains with signaling constraints is reviewed.

2.2.1 Optimal Trajectory Planning of a Single Train

The research on optimal trajectory planning for a single train started in the 1960s. A simplified train optimal control problem was studied by Ichikawa [14], who solved the problem using Pontryagin's principle. Later on, many researchers explored this optimal control problem by applying various methods, since it has significant effects for energy saving, punctuality, and riding comfort. These methods can be grouped into two main categories [8], viz., analytical solution and numerical optimization. The aim of this section is to give an overview of the research on optimal trajectory planning. Thereby, the research reported in literature will be reviewed using these two categories.

- Analytical solution
  The train is usually modeled as a point mass in the optimal control problem. According to whether the traction and braking force is continuous or discrete, there are two kinds of models, i.e. continuous-input models and discrete-input models. The research on discrete-input models is mainly done by the SCG group of the University of South Australia [6, 15]. A type of diesel-electric locomotive is considered, the throttle of which can take only on a finite number of positions. Each position determines a constant level of power supply to the wheels. Several results, which include consideration of varying grades and speed restrictions, were presented. However, nowadays many locomotives or motor cars can provide a continuous traction and braking force making the use of continuous-input models necessary. For a continuous-input model, Howlett et al. [15] gave necessary conditions for optimality and developed efficient numerical algorithms for calculating the optimal control inputs. Furthermore, in [16, 17] a new method is proposed to calculate the optimal switching points on a track with steep gradients. Alrecht et al. [18] used perturbation analysis to prove that the optimal switching points are uniquely defined for each steep section of track and deduced that the global optimal strategy is unique. Khmelnitsky [19] described the mathematical model of the train using the kinetic energy as the state variable. In that study, the optimal control problem was solved under varying grade profile and speed restrictions of rail lines. Liu and Golovicher [3] developed an analytical approach which combined the Pontryagin's principle and some algebraic equations to obtain the optimal solution, which contains the sequence of optimal controls and the change points, for the continuous-input model.
  The optimal trajectory of an analytical solution typically contains four optimal control regimes: maximum acceleration, cruising at constant speed, coasting, and maximum deceleration. It is worth to note that the analytical methods often meet difficulties if more realistic conditions are considered that introduce complex nonlinear terms into the model equations and the constraints [20].
- Numerical optimization
  A number of advanced techniques such as fuzzy and genetic algorithms have been proposed to calculate the optimal reference trajectory for trains. Chang and Xu [21] proposed a modified differential evolution algorithm to optimally tune the

fuzzy membership functions that provide a trade-off between punctuality, riding comfort, and energy consumption. The implementation of a genetic algorithm to optimize the coasting regions along a line is presented by Chang and Sim [22]. Han et al. [23] also used a genetic algorithm to construct the optimal reference trajectory taking non-constant grade profile, curve, and speed limits into account. They concluded that the performance of their genetic algorithm is better than that of the analytic solution obtained by Howlett and Pudney [6] in view of energy saving.

The train optimal control problem was solved by nonlinear programming and dynamic programming in [8]. The performance of a sequential quadratic programming algorithm and discrete dynamic programming were evaluated. Ko et al. [20] applied Bellman's dynamic programming to optimize the optimal reference trajectory. Multi-parametric quadratic programming was used to calculate optimal control laws for trains in [24]. The nonlinear train model with quadratic resistance was approximated by a piecewise affine function. The resulting optimal control law was a piecewise affine function, which relates the traction force to the train position and speed.

Due to the comparable high computing power available nowadays, more and more researchers are applying numerical optimization approaches to the train optimal control problem. However, a disadvantage of numerical solution methods is that the optimal solution is not always guaranteed and the convergence speed is uncertain in general.

2.2.2 *Optimal Trajectory Planning of Multiple Trains*

The solution approaches for the trajectory planning of a single train presented in Sect. 2.2.1 ignore the impact caused by signaling systems, e.g., an FBS system or an MBS system. In the literature, Lu and Feng [25] considered the operation of two trains on a same line and optimized the trajectory of the following train considering the constraints caused by the leading train in an FBS system. More specifically, a parallel genetic algorithm was used to optimize the trajectories for the leading train and the following train, resulting in a lower energy consumption [25]. Gu et al. [26] also considered the trajectory planning of two trains and they applied nonlinear programming to optimize the trajectory for the following train, where two situations of the leading train, i.e. running and stopped, were considered. In addition, Ding et al. [27] took the constraints caused by an MBS system into account and developed an energy-efficient multi-train control algorithm to calculate the optimal trajectories. Three optimal control regimes, i.e. maximum traction, coasting, and maximum braking, were adopted in the algorithm and the sequences of these three regimes were determined by a predefined logic [27].

For optimal trajectory planning of trains, the analytical methods often meet difficulties to find analytical solutions if more realistic conditions are considered that introduce complex nonlinear terms into the model equations and the constraints.

The numerical optimization approaches are used more and more with the increasing computing power even though the optimal solution is not always guaranteed. In Chaps. 3 and 4 of this book, we will develop efficient approaches to provide a balanced trade-off between accuracy and computational efficiency for the trajectory planning of trains. Furthermore, since the operation of trains is highly influenced by signaling systems and only a few researchers studied the impact of signaling systems in trajectory planning problem, we will also investigate the trajectory planning problem with signaling constraints in this book.

2.3 Urban Rail Transit Scheduling Process

A general scheduling both for interurban and urban rail transit systems is a highly complex process, which is often divided into several steps [28]: demand analysis, line planning, train scheduling, rolling stock planning, and crew scheduling as shown in Fig. 2.5. First, the passenger demand has to be assessed and analyzed. Consequently, the amount of travelers wishing to go from certain origins to destinations is determined. Next, line planning is performed, which decides the routes or lines to be operated and the nominal frequency of the service. During the train scheduling step, all departure and arrival times at all stations of the lines are planned, i.e., the timetable is determined. The rolling stock planning assigns trains to all the lines. Similarly, the crews are distributed to different trains through the crew scheduling. If the crew schedule is not feasible, then the rolling stock plan or even the train schedule should be adapted. In addition, some researchers have studied the integration of steps, e.g., IVU plan introduces the integration of rolling stocking planning and crew scheduling to achieve optimal deployment of resources [29].

For urban rail transit systems, not all steps are equally important. There are specific characteristics for urban rail transit systems. The degree of freedom in the line planning is limited because the routes for the operation of trains have been fixed

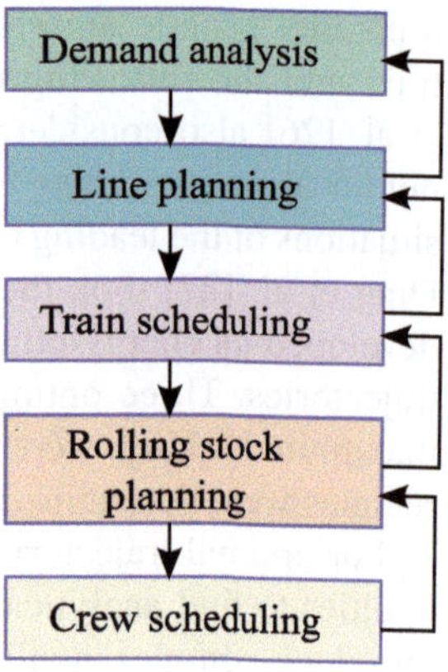

Fig. 2.5 The hierarchical planning process of railway system [28]

when the urban rail lines were constructed, i.e., trains do not move from one line to another during regular operation. Therefore, only the frequencies of the service, the stop-skip schedule on a certain line, and the size of train fleet can be regulated through coupling or decoupling of multiple train units to adapt varying passenger demands in urban rail transit lines. Therefore, in this section, the passenger demand and the train scheduling for urban rail transit systems will be discussed in detail.

2.3.1 *Passenger Demand*

Passenger demand estimation is the basis for the whole planning process. Traditionally, demand estimation relies heavily on costly and unreliable manual data collection, e.g., using passenger surveys to estimate origin-destination (OD) travel patterns. The results obtained by this kind of manual data collection maybe subject to bias and even error [30]. However, nowadays most urban rail transit systems have been equipped with automatic passenger counting systems and automatic fare collection systems, which can provide accurate passenger information to rail operators. Automatic passenger counting systems are used to count the number of passengers getting on and getting off trains at stations. With automatic fare collection systems, passengers need to use their fare cards when entering and exiting urban rail transit systems, so the location and time of each passenger's fare transactions can be recorded.

Due to historical reasons and the complexity of optimization models, the passenger demand is usually described in one of the following two ways in the literature:

- OD-independent passenger demands
  When describing the passenger demand in an OD-independent way, the origin and destination of each passenger are not considered. The passenger arrival rate at a certain station is then e.g. defined as the number of passengers arriving at the station during a predefined time period [31].
- OD-dependent passenger demands
  The OD-dependent passenger demand is defined as an estimation of the number of people wishing to travel from an origin to a destination over a certain period of time during the day. The OD-dependent passenger demand can be conducted using the available passenger information, see [30, 32, 33] for details.

2.3.2 *Train Scheduling*

Train scheduling for interurban rail transit systems has been studied for decades via different mathematical techniques [34], such as linear programming [35, 36], integer or nonlinear programming [28, 37–39], and graph theory [40]. In interurban rail transit systems, the available resources, e.g., the single tracks and the crossings, are shared by trains with different origins and destinations. Thus, the trains may overtake and cross each other at some specific locations, such as sidings and crossings. However, urban rail transit systems have the following characteristics: (1) the urban rail transit lines are usually separate from each other and have double tracks, where each track is used for one direction of the train operations, (2) train overtakings and crossings are normally not allowed during the operations, (3) the frequency of train services is much higher when compared with interurban rail transit systems. Here, we concentrate on urban rail transit systems.

2.3.2.1 Scheduling of Trains for Urban Rail Transit

In 1980, Cury et al. [41] presented a methodology to generate optimal schedules for metro lines based on a model of the train movements and of the passenger behavior. The performance index included passenger delay, passenger comfort, and the efficiency of the operation of trains. The resulting nonlinear scheduling problem was recast into several subproblems by Lagrangian relaxation and then solved in a hierarchical manner [41]. Since the convergence rate of the hierarchical decomposition algorithm can be quite poor in some cases, Assis and Milani [42] proposed a model predictive control algorithm based on linear programming to optimize the train schedule. The algorithm proposed in [42] can effectively generate train schedules for the whole day. Kwan and Chang [43] applied a heuristic-based evolutionary algorithm to solve the train scheduling problem, where the operation costs and the passenger dissatisfaction are included in the performance index. The train scheduling problem is formulated as a periodic event-scheduling problem based on a graph model in [44], which is then solved using integer programming methods. The approach proposed by Liebchen has been applied in Berlin subway systems [45]. The passenger transfer behavior and transfer waiting times are considered in [46], which presents a mixed-integer programming optimization model to synchronize the train schedules for different urban rail transit lines. Furthermore, a demand-oriented timetable design is proposed in [47], where the optimal train frequency and the capacity of trains are first determined and then the schedule of trains are optimized. Vazquez et al. [48] proposed a stochastic approximation approach to adjust the frequencies of different urban transit lines according to the observed variable passenger demand. However, the energy consumption of railway operation and dwell times at stations are not included in the model of [48].

2.3.2.2 Real-Time Scheduling or Rescheduling of Trains

Since trains do not run exactly according to the predefined schedule in practice, real-time scheduling approaches have been proposed. In the literature, there are several interpretations for real-time scheduling. For interurban railway systems, real-time scheduling is based on the existing timetable data and is used to handle route conflicts due to train delays or incidents [40, 49–56]. However, in urban rail transit systems, real-time scheduling regulates the headways between trains based on a train schedule with a constant headway.

Several rescheduling approaches have been proposed for urban rail transit systems [57]: holding, zone scheduling, short turning, deadheading, and/or stop-skipping [31, 58–60]. Holding is used to regulate the headways by holding an early-arriving train, or a train with a relatively short leading headway [31]. In zone scheduling [58], the whole line is divided into several zones, where the trains stop at all stations within a single zone and then run to the terminal station without stopping. The required number of trains and drivers and passenger travel times may be reduced by the zone scheduling, where the zones are defined based on the passenger flows. There are short-turning and full-length trips operating on the line in the short-turning strategy [59, 60], where the short-turning trips serve only the zone with high demands and the full-length trips run the whole line. The deadheading strategy involves some trains running empty through a number of stations at the beginning of their trips to reduce the headways at later stations [57, 61]. A dynamic stop-skipping strategy is frequently used in lines with high demands, as it allows those trains that are late and behind the schedule to skip certain low-demand stations and in that way increase the running speed.

Wong and Ho [62] proposed dwell time and running time control for the real-time rescheduling problem of urban rail transit systems. They applied a dynamic programming approach to their rescheduling model to devise an optimal set of dwell times and running times [62]. In addition, Goodman and Murata [63] formulated the train rescheduling problem from the perspective of passengers, where a gradient calculation method was developed to solve the rescheduling problem in real time. Furthermore, Norio et al. [64] proposed to use passenger dissatisfaction as a criterion for the rescheduling and applied a meta-heuristics algorithm to solve the rescheduling problem.

As demonstrated in [65, 66], the stop-skipping strategy can reduce the passenger travel time and the operation cost of rail transit operators. The stop-skipping operation was first developed for the Chicago metro system in 1947 [65]. Now, the SEPTA line in Philadelphia, Helsinki commuter rail, and the metro system in Santiago, Chile apply the stop-skipping train schedule in practice. They apply a static stop-skipping strategy [66], i.e., the A/B skip-stop strategy, where stations are divided into three types: A, B, and AB; A train services stop at A stations and AB stations, while B train services stop at B stations and AB stations. Major stations are usually labeled with the type AB; so all trains stop there. The transit operators provide the stop-skipping information to passengers via panels at platforms and announcements in the trains. The Santiago metro operator stated that passengers adapt to the stop-skipping strategy quickly [65]. Elberlein [67] formulated the stop-skipping problem as a mixed integer nonlinear programming problem (MINLP), where trains can skip some station strings (i.e., a collection of consecutive stations). Fu et al. [57] represented the skipping of stations by trains as binary variables and obtained a MINLP problem, which was solved using an exhaustive approach. Lee [66] applied genetic algorithm to obtain the optimal train schedule and to find the best combination of the stop-skipping trains and the all-stop trains based on the A/B stop-skipping strategy.

The passenger demand for urban rail transit systems increases dramatically and varies significantly along urban rail transit lines and the time of the day. To satisfy the passenger demand, trains are operated with small headway, which is around 2–5 min. Therefore, the scheduling of trains according to the passenger demand becomes more and more important for reducing the operation costs and for guaranteeing passenger satisfaction. In particular, the passenger satisfaction can be characterized by waiting times at platforms, onboard travel times, the number of transfers, the onboard crowdedness, etc.

2.4 Summary

A brief introduction to the operation of trains and the principle of signaling systems has been presented in this chapter. We have briefly discussed the literature of the optimal trajectory planning for trains and of the train scheduling for urban rail transit systems. In addition, we have motivated why the work of this book is needed.

References

1. Hansen I, Pachl J (2008) Railway, timetable & traffic: analysis, modelling, simulation. Eurailpress, Hamburg
2. Midya S, Thottappillil R (2008) An overview of electromagnetic compatibility challenges in European rail traffic management system. Transp Res Part C Emerg Technol 16:515–534
3. Liu R, Golovicher I (2003) Energy-efficient operation of rail vehicles. Transp Res Part A Policy Pract 37:917–931
4. Peng H (2008) Urban rail transit system. China Communication Press, Beijing
5. Dong H, Ning B, Cai B, Hou Z (2010) Automatic train control system development and simulation for high-speed railways. IEEE Circuits Syst Mag 10:6–18
6. Howlett P, Pudney P (1995) Energy-efficient train control., Advances in industrial control-Springer, London
7. Pontryagin L (1962) The mathematical theory of optimal processes. Interscience publisher, New York
8. Franke R, Meyer M, Terwiesch P (2002) Optimal control of the driving of trains. Automatisierungstechnik 50:606–614
9. Rahn K, Bode C, Albrecht T (2013) Energy-efficient driving in the context of a communications-based train control system (cbtc). In: Proceedings of the 1st IEEE international conference on intelligent rail transportation (2013 IEEE ICIRT). Beijing, China. Paper 152
10. Mitchell I (2009) The sustainable railway: use of advisory systems for energy savings. IRSE News 151:2–7
11. Pachl J (2009) Railway operation and control, 2nd edn. Gorham Printing, Centralia
12. Takeuchi H, Goodman C, Sone S (2003) Moving block signaling dynamics: performance measures and re-starting queued electric trains. IEEE Proc Electr Power Appl 150:483–492
13. Pearson L (1973) Moving block signalling. Ph.D. thesis, Loughborough University of Technology, Loughborough, UK
14. Ichikawa K (1968) Application of optimization theory for bounded state variable problems to the operation of a train. Bull Jpn Soc Mech Eng 11:857–865
15. Howlett P (2000) The optimal control of a train. Ann Oper Re 98:65–87

16. Howlett P, Pudney P, Vu X (2009) Local energy minimization in optimal train control. Automatica 45:2692–2698
17. Vu X (2006) Analysis of necessary conditions for the optimal control of a train. Ph.D. thesis, University of South Australia, Adelaide, Australia
18. Albrecht A, Howlett P, Pudney P, Vu X (2013) Energy-efficient train control: from local convexity to global optimization and uniqueness. Automatica 49:3072–3078
19. Khmelnitsky E (2000) On an optimal control problem of train operation. IEEE Trans Autom Control 45:1257–1266
20. Ko H, Koseki T, Miyatake M (2004) Application of dynamic programming to optimization of running profile of a train. Computers in railways IX. WIT Press, Southhampton, pp 103–112
21. Chang C, Xu D (2000) Differential evolution based tuning of fuzzy automatic train operation for mass rapid transit system. IEE Proc Electr Power Appl 147:206–212
22. Chang C, Sim S (1997) Optimizing train movements through coast control using genetic algorithms. IEE Proc Electr Power Appl 144:65–73
23. Han S, Byen Y, Baek J, An T, Lee S, Park H (1999) An optimal automatic train operation (ATO) control using genetic algorithms (GA). In: Proceedings of the IEEE region 10 conference (TENCON 99). Cheju Island, South Korea, pp 360–362
24. Vašak M, Baotić M, Perić N, Bago M (2009) Optimal rail route energy management under constraints and fixed arrival time. In: Proceedings of the European control conference. Budapest, Hungary, pp 2972–2977
25. Lu Q, Feng X (2011) Optimal control strategy for energy saving in trains under the four-aspected fixed auto block system. J Mod Transp 19:82–87
26. Gu Q, Lu X, Tang T (2011) Energy saving for automatic train control in moving block signaling system. In: Proceedings of the 14th international IEEE conference on intelligent transportation systems. DC, USA, Washington, pp 1305–1310
27. Ding Y, Bai Y, Liu F, Mao B (2009) Simulation algorithm for energy-efficient train control under moving block system. In: Proceedings of the 2009 world congress on computer science and information engineering. Los Angeles, CA, USA, pp 498–502
28. Ghoseiri K, Szidarovszky F, Asgharpour M (2004) A multi-objective train scheduling model and solution. Transp Res Part B 38:927–952
29. IVU.rail (2015) Integrated scheduling, dispatching and optimization
30. Zhao J, Rahbee A, Wilson NHM (2007) Estimating a rail passenger trip origin-destination matrix using automatic data collection systems. Comput Aided Civ Infrastruct Eng 22:376–387
31. Elberlein X, Wilson N, Bernstein D (2001) The holding problem with real-time information available. Transp Sci 35:1–18
32. Wong S, Tong C (1998) Estimation of time-dependent origindestination matrices for transit networks. Transp Res Part B 32:35–48
33. Li Y, Cassidy M (2007) A generalized and efficient algorithm for estimating transit route ODs from passenger counts. Transp Res Part B 41:114–125
34. Cordeau J, Toth P, Vigo D (1998) A survey of optimization models for train routing and scheduling. Transp Sci 32:380–420
35. Szpigel B (1972) Optimal train scheduling on a single line railway. In: Proceedings of the international conference on operational research. The Netherlands, Amsterdam, pp 344–351
36. Petersen E, Taylor A, Martland C (1986) An introduction to computer-assisted train dispatch. J Adv Transp 20:63–72
37. Kraay D, Harker P, Chen B (1991) Optimal pacing of trains in freight railroads: model formulation and solution. Oper Res 39:82–99
38. Higgins A, Kozan E, Ferreira L (1996) Optimal scheduling of trains on a single line track. Transp Res Part B 30:147–161
39. Li X, Wang D, Li K, Gao Z (2013) A green train scheduling model and fuzzy multi-objective optimization algorithm. Appl Math Model 37:617–629
40. D'Ariano A, Pranzoand M, Hansen I (2007) Conflict resolution and train speed coordination for solving real-time timetable perturbations. IEEE Trans Intell Transp Syst 8:208–222

41. Cury J, Gomide F, Mendes MJ (1980) A methodology for generation of optimal schedules for an underground railway systems. IEEE Trans Autom Control 25:217–222
42. Assis W, Milani B (2004) Generation of optimal schedules for metro lines using model predictive control. Automatica 40:1397–1404
43. Kwan C, Chang C (2005) Application of evolutionary algorithm on a transportation scheduling problem—the mass rapid transit. In: Proceedings of the IEEE congress on evolutionary computation. Edinburgh, UK, pp 987–994
44. Liebchen C (2006) Periodic timetable optimization in public Transport. Ph.D. thesis, Technique University of Berlin, Berlin, Germany
45. Liebchen C (2008) The first optimized railway timetable in practice. Transp Sci 42:420–435
46. Wong R, Yuen T, Fung K, Leung J (2008) Optimizing timetable synchronization for rail mass transit. Transp Sci 42:57–69
47. Albrecht T (2009) Automated timetable design for demand-oriented service on suburban railways. Public Transp 1:5–20
48. Vázquez-Abad F, Zubieta L (2005) Ghost simulation model for the optimization of an urban subway system. Discrete Event Dyn Syst 15:207–235
49. Corman F, D'Ariano A, Pacciarelli D, Pranzo M (2012) Bi-objective conflict detection and resolution in railway traffic management. Transp Res Part C Emerg Technol 20:79–94
50. Krasemann JT (2012) Design of an effective algorithm for fast response to the re-scheduling of railway traffic during disturbances. Transp Res Part C Emerg Technol 20:62–78
51. Dollevoet T, Corman F, D'Ariano A, Huisman D (2012) An iterative optimization framework for delay management and train scheduling. Technical Report EI 2012–10. Erasmus School of Economics (ESE), Rotterdam, The Netherlands
52. Cacchiani V, Toth P (2009) Nominal and robust train timetabling problems. Eur J Oper Res 219:727–737
53. Khan M, Zhou X (2010) Stochastic optimization model and solution algorithm for robust double-track train-timetabling problem. IEEE Trans Intell Transp Syst 11:81–89
54. Meng L, Zhou X (2011) Robust single-track train dispatching model under a dynamic and stochastic environment: A scenario-based rolling horizon solution approach. Transp Res Part B Methodological 45:1080–1102
55. Vansteenwegen P, Oudheusden DV (2006) Developing railway timetables which guarantee a better service. Eur J Oper Res 173:337–350
56. Kersbergen B, van den Boom T, De Schutter B (2013) Reducing the time needed to solve the global rescheduling problem for railway networks. In: Proceedings of the 16th international IEEE conference on intelligent transportation systems (ITSC 2013). The Netherlands, The Hague, pp 791–796
57. Fu L, Liu Q, Calamai P (2003) Real-time optimization model for dynamic scheduling of transit operations. Transp Res Rec 1857:48–55
58. Ghoneim N, Wirasinghe S (1986) Optimum zone structure during peak periods for existing urban rail lines. Transp Res Part B 20:7–18
59. Ceder A (1989) Optimal design of transit short-turn trips. Transp Res Rec 1221:8–22
60. Site P, Filippi F (1998) Service optimization for bus corridors with short-turn strategies and variable vehicle size. Transp Res Part A 32:19–38
61. Elberlein X, Wilson N, Barnhart C (1998) The real-time deadheading problem in transit operation control. Transp Res Part B 32:77–100
62. Wong K, Ho T (2007) Dwell-time and run-time control for dc mass rapid transit railways. IET Electr Power Appl 1:956–966
63. Goodman C, Murata S (2001) Metro traffic regulation from the passenger perspective. Proc Inst Mech Eng Part F: J Rail Rapid Transit 215:137–147
64. Norio T, Yoshiaki T, Noriyuki T, Chikara H, Kunimitsu M (2005) Train rescheduling algorithm which minimizes passengers' dissatisfaction. Lecture notes in computer science: innovations in applied artificial intelligence. Springer, Berlin, pp 829–838
65. Freyss M, Giesen R, Muñoz JC (2013) Continuous approximation for skip-stop operation in rail transit. Transp Res Part C 36:419–433

66. Lee Y (2012) Mathematical modeling for optimizing skip-stop rail transit operation strategy using genetic algorithm. Technical Report. Morgan State University, Department of Transportation and Urban Infrastructure Studies. Baltimore, MD, USA
67. Elberlein X (1995) Real-time control strategies in transit operations: models and analysis. Ph.D. thesis, Massachusetts Institute of Technology, Cambridge, MA, USA

Chapter 3
Optimal Trajectory Planning for a Single Train

Abstract In this chapter, the optimal trajectory planning problem for the operation of a single train under various constraints and with a fixed arrival time is considered. The objective function corresponds to a trade-off between the energy consumption and the riding comfort. Two approaches are proposed to solve this optimal control problem, viz. a pseudospectral method and a mixed integer linear programming (MILP) approach. In the pseudospectral method, the optimal trajectory planning problem is recast into a multiple-phase optimal control problem, which is then transformed into a nonlinear programming problem. For the MILP approach, the optimal trajectory planning problem is reformulated as an MILP problem by approximating the nonlinear terms by piecewise affine functions. The performance of these two approaches is compared through a case study. The work discussed in this chapter is based on Wang et al. (Proceedings of the 14th international IEEE conference on intelligent transportation systems (ITSC 2011). Washington DC, USA, pp 1598–1604, 2011) [1]; Wang et al. (Transp Res Part C 29:97–114, 2013) [2]; Wang et al. (Proceedings of the 13th IFAC symposium on control in transportation systems (CTS'2012). Sofia, Bulgaria, pp 158–163, 2012) [3].

3.1 Introduction

As has been mentioned in the literature survey of Chap. 2, it is important to design efficient algorithms to find the optimal speed–position reference trajectory. This chapter proposes two approaches to determine the optimal trajectory, viz. a pseudospectral method and an mixed integer linear programming (MILP) approach [1–3].

Over the last decade, pseudospectral methods have risen to prominence in the numerical optimal control area [4]. Pseudospectral methods were applied to solving optimal control problems [5], such as orbit transfers, lunar guidance, and magnetic control. However, to the author's best knowledge, pseudospectral methods have not been applied to trajectory planning of trains. Therefore, the pseudospectral method is used for the first time to solve the train trajectory planning problem.

On the other hand, multiparametric quadratic programming is used in [6] to calculate the optimal control law for train operations. The nonlinear train model with

Y. Wang et al., *Optimal Trajectory Planning and Train Scheduling for Urban Rail Transit Systems*, Advances in Industrial Control,
DOI 10.1007/978-3-319-30889-0_3

quadratic resistance is approximated by a piecewise affine function. Inspired by [6], in this chapter we propose to solve the optimal trajectory problem as an MILP problem.

The remainder of this chapter is organized as follows. In Sect. 3.2 a nonlinear model of train operations is presented. Section 3.3 formulates the optimal trajectory planning problem. In Sect. 3.4, two solution approaches are proposed to solve the resulting optimal control problem: a pseudospectral method and a MILP approach. Section 3.5 illustrates with a case study how to calculate the optimal reference trajectory by the pseudospectral method and the MILP approach and it also compares these two approaches with the discrete dynamic programming (DDP) approach. We conclude with a short discussion of some topics for future work in Sect. 3.6.

3.2 Model Formulation

3.2.1 Train Model

In the literature on train optimal control, the mass-point model of train is often adopted [7, 8], so the operation of a train can be described by a simple continuous-time model given as follows:

$$m\rho\frac{\mathrm{d}v}{\mathrm{d}t} = u(t) - R_{\mathrm{b}}(v) - R_{\mathrm{l}}(s, v), \tag{3.1}$$

$$\frac{\mathrm{d}s}{\mathrm{d}t} = v, \tag{3.2}$$

where m is the mass of the train, ρ is the rotating mass factor [9], v is the velocity of the train, s is the position of the train, u is the traction or braking force, $R_{\mathrm{b}}(v)$ is the basic resistance including roll resistance and air resistance, and $R_{\mathrm{l}}(s, v)$ is the line resistance caused by track grade, curves, and tunnels.

The traction or braking force is bounded by the maximum traction force $u_{\max}^{\mathrm{acc}}$ and the maximum braking force $u_{\max}^{\mathrm{dec}}$ with $u_{\max}^{\mathrm{dec}} > 0$, i.e., $-u_{\max}^{\mathrm{dec}} \le u \le u_{\max}^{\mathrm{acc}}$. Furthermore, the maximum traction force $u_{\max}^{\mathrm{acc}}$ is not a constant in practice, but is a function of the velocity v because of the maximum adhesion and the characteristics of the power equipment. A typical diagram of the maximum traction force $u_{\max}^{\mathrm{acc}}$ is presented in Fig. 3.1 [9], where the nonlinear function can be described as a group of hyperbolic or parabolic formulas as explained in [9]. More specifically, each hyperbolic or parabolic formula approximates the actual traction force for a certain speed interval. For example, if the train speed v belongs to interval $[v_j, v_{j+1}]$, then the maximum traction force can be formulated as

$$u_{\max}^{\mathrm{acc}}(v) = a_{1,j} + a_{2,j}v + a_{3,j}v^2, \quad v \in [v_j, v_{j+1}], \tag{3.3}$$

or

$$u_{\max}^{\mathrm{acc}}(v) = a_{\mathrm{h},j}/v, \quad v \in [v_j, v_{j+1}], \tag{3.4}$$

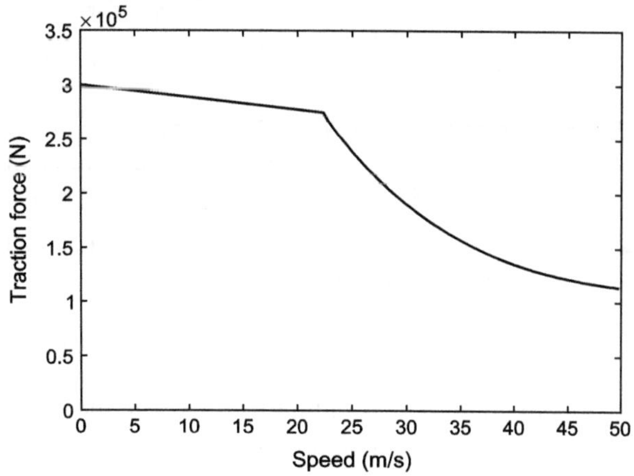

Fig. 3.1 Maximum traction force as a function of train speed [9]

for $j = 1, 2, \ldots, J-1$, where J, v_j, v_{j+1}, $a_{1,j}$, $a_{2,j}$, $a_{3,j}$, and $a_{h,j}$ are determined by the characteristics of the train.

Based on the arguments about the maximum braking force stated in [9], the full braking effort is only reserved for an emergency stop. Under normal circumstances (i.e., normal braking processes) the train driver or automatic train operation (ATO) system brakes in a more comfortable way, where the maximum service braking force is three quarters of the full braking force in general. In addition, the maximum service braking force (including the full braking force) is normally considered as a constant by some common safety systems, such as the European train control system and the German continuous train control system [9]. Hence, the maximum force for service braking is also taken to be a constant in this chapter.

According to the Strahl formula [10], in practice the basic resistance $R_b(v)$ can be described by

$$R_b(v) = m(a_1 + a_2 v^2),$$

where the coefficients a_1 and a_2 depend on the train characteristics, the shape of the train's head, the wind speed, etc. These coefficients can be estimated from the operation test data of the train.

The line resistance $R_l(s, v)$ induced by track slope, curves, and tunnels can be formulated as [11]

$$R_l(s, v) = mg \sin \alpha(s) + f_c(r(s)) + f_t(l_t(s), v), \qquad (3.5)$$

where g is the gravitational acceleration, $\alpha(s)$, $r(s)$, and $l_t(s)$ are the slope, the radius of the curve, and the length of the tunnel along the track, respectively. The curve resistance $f_c(\cdot)$ and the tunnel resistance $f_t(\cdot)$ are described by empirical formulas. For example, Roeckl's formula [12] is an empirical formula for the curve resistance, which is given as follows:

$$f_c(r(s)) = \begin{cases} \frac{6.3}{r(s)-55} m & \text{for} \quad r(s) \geq 300\,\text{m}, \\ \frac{4.91}{r(s)-30} m & \text{for} \quad r(s) < 300\,\text{m}. \end{cases}$$

When the train running in tunnels, it experiences a higher air resistance that depends on the tunnel form, the smoothness of tunnel walls, the exterior surface of the train, etc. An example of an expression for the tunnel resistance is as follows [12, 13]:

- If there exists a limiting gradient[1] in the tunnel, then an empirical formula for the tunnel resistance can be computed by

$$f_t(l_t(s), v) = 1.296 \times 10^{-9} l_t(s) m g v^2.$$

- If there does not exist a limiting gradient, the tunnel resistance can be calculated by the following empirical formula:

$$f_t(l_t(s), v) = 1.3 \times 10^{-7} l_t(s) m g.$$

For the tracks outside the tunnels, the tunnel resistance is equal to zero.

Remark 3.1 Various types of rolling stock can be described by the mass-point model, the parameters of which, such as mass, maximum traction force, and resistance coefficients, may be different for each type of rolling stock. The existing infrastructure of tracks can be described precisely by the line resistance (3.5), which involves the track slope, curves, tunnels, etc. Moreover, the disturbances caused by other trains and the signaling systems are assumed to be managed by a lower control level. Furthermore, different train categories (high speed trains, intercity trains, regional trains, and freight trains) can be handled by the train scheduling, which specifies the running times and dwell times for each train. The approaches presented in this chapter can then be used to calculate the optimal speed profile for each trip between two stations to save energy and to ensure passenger comfort according to the given timetable.

3.2.2 An Assumption About the Line Resistance

The line resistance $R_l(\cdot, \cdot)$ induced by track slope, curves, and tunnels is a nonlinear function of the train's position s and speed v. In order to simplify the formulation of the line resistance, we rewrite $R_l(s, v)$ in (3.5) as

$$R_l(s, v) = \xi_1(s) + \xi_2(s) v^2, \tag{3.6}$$

where $\xi_1(s)$ collects the terms that do not rely on the train's speed. In the sequel of this chapter, $\xi_1(\cdot)$ and $\xi_2(\cdot)$ are assumed to be piecewise constant functions, which can be written as

[1] A limiting gradient is defined as the maximum railway gradient that can be climbed without the help of a second power unit.

$$\xi_1(s) = \xi_1^{(i)} \quad \text{for} \quad s \in [s_0^{(i)}, s_f^{(i)}],$$
$$\xi_2(s) = \xi_2^{(i)} \quad \text{for} \quad s \in [s_0^{(i)}, s_f^{(i)}], \tag{3.7}$$

for $i = 1, 2, \ldots, N_R$, where N_R is the number of the piecewise constant subfunctions, $s_0^{(1)} = s_{\text{start}}$ is the position at the beginning of the route, $s_f^{(N_R)} = s_{\text{end}}$ is the position at the end of the route, and $s_0^{(i+1)} = s_f^{(i)}$ for $i = 1, 2, \ldots, N_R - 1$. Hence, the line resistance can be written as

$$R_l(s, v) = \xi_1^{(i)} + \xi_2^{(i)} v^2, \quad \text{for} \quad s \in [s_0^{(i)}, s_f^{(i)}]. \tag{3.8}$$

3.3 Mathematical Formulation of the Single Train Trajectory Planning Problem

The optimal trajectory planning for trains can be formulated as an optimal control problem as stated in [8], where the control variable is the traction or braking force u, the state variables are the train position s and speed v, and the objective function to be minimized could be the trip time, the energy consumption for a given trip time, or the total operation cost (a weighted sum of energy consumption and trip time). In this chapter, we take the objective criterion as the energy consumption in a fixed time span $[0, T]$ where T is determined by a given timetable [9, 14], or is obtained by the rescheduling process of railway traffic in case of disturbances [15]. Moreover, the riding comfort is also taken into account in the objective criterion. The riding comfort can be expressed as a function of the change of the control variable u, since reducing the number of transitions and the rate of change of u will improve passenger comfort [16]. The objective function can thus be written as:

$$J = \int_0^T \left(u(t) \cdot v(t) + \lambda \cdot \left| \frac{\mathrm{d}u(t)}{\mathrm{d}t} \right| \right) \mathrm{d}t, \tag{3.9}$$

where J is a weighted integral of the energy consumption and riding comfort and $\lambda > 0$ is the weight which can be determined by rail operators on the basis of operation experience. A weight decision structure as that proposed in [17] for the ATO system could be adopted by the rail operators. This function will be minimized subject to the train dynamics (3.1) and (3.2), the constraints

$$-u_{\max}^{\text{dec}} \le u(t) \le u_{\max}^{\text{acc}}(v), \tag{3.10}$$

$$0 \le v(t) \le V_{\max}(s), \tag{3.11}$$

and the boundary conditions

$$s(0) = s_{\text{start}}, \quad v(0) = v_{\text{start}}, \tag{3.12}$$

$$s(T) = s_{\text{end}}, \; v(T) = v_{\text{end}}, \tag{3.13}$$

where the maximum allowable velocity $V_{\max}(s)$ is up to the train characteristics and the line conditions, and as such it is usually a piecewise constant function of the coordinate s [8, 18]; v_{start} and v_{end} are the velocity at the beginning and the end of the route.

As proposed in some previous works [7, 8, 18, 19], it is better to take the position s as the independent variable rather than the time t, because the consideration of track-related data, such as line resistance and speed limits, will be simplified by taking the position s as the independent variable. Moreover, the analytical and numerical study of the optimal control problem will be significantly simplified. In addition, Khmelnitsky [18] chose the total energy of the train and time t as states where the total energy includes kinetic and potential energy. Similarly, Franke et al. [7] employed kinetic energy per mass unit and time as states. The choice of kinetic energy instead of speed v will facilitate the study of the optimal control problem, because some (but not all) of the model nonlinearities will be eliminated by this choice. As a result, we also choose kinetic energy per mass unit $\tilde{E} = 0.5v^2$ and time t as states, and the position s as the independent variable. The continuous-time model (3.1) and (3.2) can then be rewritten as the following continuous-space model[2]:

$$m\rho\frac{\mathrm{d}\tilde{E}}{\mathrm{d}s} = u(s) - R_{\text{b}}(\sqrt{2\tilde{E}}) - R_{\text{l}}(s, v), \tag{3.14}$$

$$\frac{\mathrm{d}t}{\mathrm{d}s} = \frac{1}{\sqrt{2\tilde{E}}}. \tag{3.15}$$

The optimal control problem described in (3.9)–(3.13) can be rewritten as: minimizing the objective function[3]

$$J = \int_{s_{\text{start}}}^{s_{\text{end}}} \left(u(s) + \lambda \cdot \left|\frac{\mathrm{d}u(s)}{\mathrm{d}s}\right|\right)\mathrm{d}s, \tag{3.16}$$

---

[2]The transformation from $\frac{\mathrm{d}v}{\mathrm{d}t}$ to $\frac{\mathrm{d}\tilde{E}}{\mathrm{d}s}$ goes as follows:

$$\frac{\mathrm{d}v}{\mathrm{d}t} = \frac{\mathrm{d}v}{\mathrm{d}s}\frac{\mathrm{d}s}{\mathrm{d}t} = v\frac{\mathrm{d}v}{\mathrm{d}s} = \frac{\mathrm{d}\tilde{E}}{\mathrm{d}s},$$

where $\tilde{E} = 0.5v^2$.

[3]The transformation from $uv\mathrm{d}t$ to $u\mathrm{d}s$ goes as follows:

$$u \cdot v\,\mathrm{d}t = u\frac{\mathrm{d}s}{\mathrm{d}t}\mathrm{d}t = u\,\mathrm{d}s.$$

In addition, the transformation from $\left|\frac{\mathrm{d}u}{\mathrm{d}t}\right|\mathrm{d}t$ to $\left|\frac{\mathrm{d}u}{\mathrm{d}s}\right|\mathrm{d}s$ goes as follows:

$$\left|\frac{\mathrm{d}u}{\mathrm{d}t}\right|\mathrm{d}t = \left|\frac{\mathrm{d}u}{\mathrm{d}s}\right|\left|\frac{\mathrm{d}s}{\mathrm{d}t}\right|\mathrm{d}t = \left|\frac{\mathrm{d}u}{\mathrm{d}s}\right|\mathrm{d}s, \text{ if } \frac{\mathrm{d}s}{\mathrm{d}t} > 0.$$

subject to the model (3.14) and (3.15), the constraints

$$-u_{\max}^{\mathrm{dec}} \leq u(s) \leq u_{\max}^{\mathrm{acc}}(v), \tag{3.17}$$

$$0 \leq \tilde{E}(s) \leq \tilde{E}_{\max}(s), \tag{3.18}$$

and the boundary conditions,

$$\tilde{E}(s_{\mathrm{start}}) = \tilde{E}_{\mathrm{start}}, \quad \tilde{E}(s_{\mathrm{end}}) = \tilde{E}_{\mathrm{end}}, \tag{3.19}$$

$$t(s_{\mathrm{start}}) = 0, \quad t(s_{\mathrm{end}}) = T, \tag{3.20}$$

where $\tilde{E}_{\max}(s) = 0.5V_{\max}^2(s)$, $\tilde{E}_{\mathrm{start}} = 0.5v_{\mathrm{start}}^2$, and $\tilde{E}_{\mathrm{end}} = 0.5v_{\mathrm{end}}^2$. As for the above equations, it is assumed that the unit kinetic energy $\tilde{E}(s)$ satisfies $\tilde{E}(s) \geq E_{\min} > 0$ with $E_{\min}$ a small positive number, which means the train's speed is always strictly larger than zero, i.e., the train travels with nonstop. As stated by Khmelnitsky [18], traveling with nonstop is not restrictive in practice for the trajectory planning because the speed of the start and the end point can be approximated by small nonzero velocities. Moreover, intermediate stops during the trip will in principle not be planned deliberately in the optimal control design for a single train's operation since intermediate stops and the corresponding restarts would lead to an increased energy consumption.

Remark 3.2 There could be other constraints caused by the timetable, real-time operation restrictions, or the real-time rescheduling process. Albrecht et al. [20, 21] classified the operational constraints into two groups: target points and target windows. Target points correspond to fixed passing times, which could be arrival and departure times at stations. In dense networks, target points could also be passing times at certain places where overtaking and crossing of trains is planned. The scheduled arrival times at minor stations without connections with other trains can be in general regarded as target windows. If the train reaches a certain place exactly on time according to the defined target point or in the target window, then conflicts can be avoided. It is assumed that the positions corresponding to target points or target window constraints are s_{k_j} with k_j in $\{1, 2, \ldots, N\}$. The operational constraints can be included in the optimal control problem as follows:

- For target points:

$$t(s_{k_j}) = T_{\mathrm{target},j}, \tag{3.21}$$

- For target windows:

$$T_{\mathrm{target}_{\min},j} \leq t(s_{k_j}) \leq T_{\mathrm{target}_{\max},j}, \tag{3.22}$$

where $T_{\text{target},j}$ is the fixed passing time for train to pass position s_{k_j}, and $T_{\text{target}_{\min},j}$ and $T_{\text{target}_{\max},j}$ are the minimum and maximum passing time at position s_{k_j} respect to the target window constraints.

Remark 3.3 In order to deal with disturbance, we can apply a rolling horizon approach when solving the trajectory planning problem. A detailed description on how to do this for the case of train scheduling for an urban rail transit network is given in Sect. 7.4. Those ideas can be adopted to the setting of the current chapter.

3.4 Solution Approaches

In this section, the train trajectory planning problem is solved by a pseudospectral method and a MILP approach.

3.4.1 Pseudospectral Method

3.4.1.1 A Brief Introduction

Pseudospectral methods were researched widely in the 1970s for solving partial differential equations (PDEs) in fluid dynamics [22]. Later on, they became an important methodology for the numerical solution of PDEs. From the 1990s on, pseudospectral methods were applied for solving optimal control problems [5], such as orbit transfers, lunar guidance, and magnetic control. Recently, the scope of application has been broadened as a result of significant progress in large-scale computation.

The pseudospectral method directly formulates the original optimal control problem into a nonlinear programming problem, which can be solved numerically using a sparse nonlinear programming solver to obtain approximate locally optimal solutions [4]. It is shown by approximation theory and practice that the pseudospectral method is well-suited for approximating smooth functions, integrations, and differentiations [22]. All those approximations are related to optimal control problems, e.g., the differential equations of the optimal control problem can then be approximated by algebraic equations [23]. The main advantages of the pseudospectral method are the exponential rate of convergence and that it is possible to achieve a good accuracy with coarse grids [22, 24].

In the pseudospectral method, the continuous-time state and control functions are approximated by orthogonal polynomials based on interpolation at orthogonal collocation points [25], such as the commonly used Legendre-Gauss-Lobatto points. Legendre-Gauss-Lobatto points are the roots of $(1-x^2)\frac{\mathrm{d}L_{N(x)}}{\mathrm{d}x}$, where L_N is the Legendre polynomial of order N [22]. The derivative of the approximated state can be expressed in terms of the approximated state vector using a differentiation matrix at the collocation points [26]. When the optimal control problem includes discontinuities in states, control inputs, objective functional, or dynamic constraints, the pseudospectral method can be applied in the form of a multiple-phase approach, where the problem is split into a relatively small number of subintervals and global collocation is performed in each subinterval [26].

There are several commercial and free packages that implement the pseudospectral method: PROPT [27] and DIDO [28] are examples of commercial software that run under MATLAB®. A MATLAB-based open source tool that employs the Gauss pseudospectral method is GPOPS [29]. PSOPT is an open source optimal control package written in C++, including Legendre and Chebyshev pseudospectral discretizations [30].

3.4.1.2 Formulation of the Optimal Trajectory Planning Problem

The train trajectory planning problem (3.14)–(3.20) can be reformulated into the following general optimal control problem with N_p phases [26, 31]. It is worth noting that N_p is not equal to N_R of (3.7) and (3.8), but it will in general be larger. The objective function (3.16) to be minimized can be rewritten as

$$J = \sum_{i=1}^{N_\mathrm{p}} \left[\int_{s_0^{(i)}}^{s_\mathrm{f}^{(i)}} \left[u^{(i)}(s) + \lambda \left| \frac{\mathrm{d}u^{(i)}(s)}{\mathrm{d}s} \right| \right] \mathrm{d}s \right]. \tag{3.23}$$

Given that non-smoothness leads to problems in a gradient-based nonlinear optimization, a smooth version of the absolute value function can be written as

$$|\sigma| \approx \psi(\sigma) = \frac{\sigma^2}{\sqrt{\sigma^2 + c^2}}, \tag{3.24}$$

where c is a constant that determines the smoothness of the function. As a result, the smooth objective function can be written as

$$J = \sum_{i=1}^{N_\mathrm{p}} \left[\int_{s_0^{(i)}}^{s_\mathrm{f}^{(i)}} \left[u^{(i)}(s) + \lambda \psi \left(\frac{\mathrm{d}u^{(i)}(s)}{\mathrm{d}s} \right) \right] \mathrm{d}s \right]. \tag{3.25}$$

The objective function (3.25) is subject to the differential constraints

$$\dot{x}^{(i)}(s) = \phi^{(i)}\left(x^{(i)}(s), u^{(i)}(s), s\right), \quad s \in [s_0^{(i)}, s_\mathrm{f}^{(i)}], \tag{3.26}$$

where $x^{(i)}(s)$ is the state of the system in the ith phase, i.e., $x^{(i)}(s) = \left[E^{(i)}(s)\ t^{(i)}(s)\right]^T$, and the functions $\phi^{(i)}(\cdot)$ are defined by model equations (3.14) and (3.15) and the piecewise line resistance (3.8). The path constraints of the optimal control problem are defined by (3.3), (3.4), and (3.17). It is noted that the path constraints caused by the maximum traction force are non-smooth. By introducing a smooth version of the Heaviside function, the path constraints can be approximated by smooth constraints, where the Heaviside function $H(\sigma)$ is defined as $H(\sigma) = 1$ if $\sigma > 0$, and $H(\sigma) = 0$ otherwise [32]. The approximation of the Heaviside function could be implemented as

$$H(\sigma) \approx 0.5(1 + \tanh(\sigma/h)), \tag{3.27}$$

where $h > 0$ is a small real number. The path constraints can then be written as

$$\mathfrak{p}_{\mathrm{L}}^{(i)} \leq \mathfrak{p}^{(i)}(x^{(i)}(s), u^{(i)}(s), s) \leq \mathfrak{p}_{\mathrm{U}}^{(i)}, \quad s \in [s_0^{(i)}, s_{\mathrm{f}}^{(i)}]. \tag{3.28}$$

For the train trajectory planning problem, the initial position of the $(i+1)$th phase is equal to the final position of the ith phase, so one of the phase boundary constraints can be written as

$$s_0^{(i+1)} - s_{\mathrm{f}}^{(i)} = 0. \tag{3.29}$$

Moreover, the states and the control variables are continuous across the phase boundary, which can be formulated as

$$x(s_0^{(i+1)}) - x(s_{\mathrm{f}}^{(i)}) = 0, \tag{3.30}$$

$$u(s_0^{(i+1)}) - u(s_{\mathrm{f}}^{(i)}) = 0. \tag{3.31}$$

Generally, phase boundary constraints [31, 33] that link all the states and control inputs across the boundaries can be written as

$$\begin{aligned}\Psi_{\mathrm{L}} \leq \Psi(&x^{(1)}(s_0^{(1)}), u^{(1)}(s_0^{(1)}), x^{(1)}(s_{\mathrm{f}}^{(1)}), u^{(1)}(s_{\mathrm{f}}^{(1)}), s_0^{(1)}, s_{\mathrm{f}}^{(1)},\\ &x^{(2)}(s_0^{(2)}), u^{(2)}(s_0^{(2)}), x^{(2)}(s_{\mathrm{f}}^{(2)}), u^{(2)}(s_{\mathrm{f}}^{(2)}), s_0^{(2)}, s_{\mathrm{f}}^{(2)},\\ &\vdots\\ &x^{(N_{\mathrm{p}})}(s_0^{(N_{\mathrm{p}})}), u^{(N_{\mathrm{p}})}(s_0^{(N_{\mathrm{p}})}), x^{(N_{\mathrm{p}})}(s_{\mathrm{f}}^{(N_{\mathrm{p}})}), u^{(N_{\mathrm{p}})}(s_{\mathrm{f}}^{(N_{\mathrm{p}})}), s_0^{(N_{\mathrm{p}})}, s_{\mathrm{f}}^{(N_{\mathrm{p}})}) \leq \Psi_{\mathrm{U}}.\end{aligned} \tag{3.32}$$

Note that (3.29)–(3.31) are special cases of (3.32) with $\Psi_{\mathrm{L},i} = \Psi_{\mathrm{U},i}$. The bound constraints can be written as

$$\begin{aligned} u_{\mathrm{L}}^{(i)} &\leq u^{(i)}(s) \leq u_{\mathrm{U}}^{(i)}, \quad s \in [s_0^{(i)}, s_{\mathrm{f}}^{(i)}],\\ x_{\mathrm{L}}^{(i)} &\leq x^{(i)}(s) \leq x_{\mathrm{U}}^{(i)}, \quad s \in [s_0^{(i)}, s_{\mathrm{f}}^{(i)}]. \end{aligned} \tag{3.33}$$

The resulting multiple-phase optimal control problem can be solved by the nonlinear programming methods.

However, the computation of the pseudospectral method is in general too slow for the real-time application of ATO system. When the operational conditions (e.g., speed limits or trip time) change during the operation of a train (e.g., due to an accident or bad weather conditions), the ATO system needs to recalculate the optimal trajectory.

If the algorithm of the ATO system takes a long time to calculate the optimal trajectory, then it is too late for the train to react. Therefore, an alternative approach, i.e., an MILP approach, is proposed in the next subsection to calculate the optimal trajectory. It is worth to note that the optimal solution of the pseudospectral method satisfies the necessary (but not always sufficient) conditions of optimality [27]. So it is guaranteed that the returned solution cannot be improved by an infinitesimal change in the trajectory, but there may exist completely different trajectories that yield a better performance. On the contrary, an MILP problem can be solved efficiently by existing solvers that guarantee the global optimum for the proposed MILP problem.

3.4.2 Mixed Integer Linear Programming

Vašak et al. [6] presented a discrete-time model of the train operation to calculate the optimal control law by multiparametric quadratic programming. They split the time period into K intervals and assumed the traction force or braking force to be constant on each interval $[kT_s, (k+1)T_s)$, where T_s is the sampling time. Franke et al. [7] similarly split the position horizon $[s_{\text{start}}, s_{\text{end}}]$ into N intervals to get a discrete-space model. They assumed that the track and train parameters as well as traction or braking force can be considered as constant in each interval $[s_k, s_{k+1}]$ with length $\Delta s_k = s_{k+1} - s_k$, for $k = 1, 2, \ldots, N$. Note that $s_1 = s_{\text{start}}$ and $s_{N+1} = s_{\text{end}}$. In this chapter, we obtain a discrete-space model in a similar way as in [7], since the optimal control problem is stated by the choice of s as the independent variable. By redefining the discretization of the interval $[s_{\text{start}}, s_{\text{end}}]$ if necessary, we can assume without loss of generality that $\xi_1(s)$ and $\xi_2(s)$ (cf. Sect. 3.2.2) are of the following form:

$$\xi_1(s) = \xi_{1,k} \quad \text{for} \quad s \in [s_k, s_{k+1}],$$
$$\xi_2(s) = \xi_{2,k} \quad \text{for} \quad s \in [s_k, s_{k+1}],$$

for $k = 1, 2, \ldots, N$.

3.4.2.1 Transformation Properties

First, we introduce three properties according to [34]. Consider the statement $\tilde{f}(\tilde{x}) \leq 0$, where $\tilde{f} : \mathbb{R}^n \to \mathbb{R}$ is affine, $\tilde{x} \in \chi$ with $\chi \subset \mathbb{R}^n$ and let

$$\tilde{M} = \max_{\tilde{x} \in \chi} \tilde{f}(\tilde{x}), \; \tilde{m} = \min_{\tilde{x} \in \chi} \tilde{f}(\tilde{x}). \tag{3.34}$$

If we introduce the logical variable $\delta \in \{0, 1\}$, then the following equivalence holds:

$$[\tilde{f}(\tilde{x}) \leq 0] \Leftrightarrow [\delta = 1] \quad \text{is true iff} \quad \begin{cases} \tilde{f}(\tilde{x}) \leq \tilde{M}(1-\delta), \\ \tilde{f}(\tilde{x}) \geq \varepsilon + (\tilde{m} - \varepsilon)\delta, \end{cases} \tag{3.35}$$

where ε is a small positive number (typically the machine precision) that is introduced to transform a strict equality into a non-strict inequality, which fits the MILP framework [34].

The product of two logical variables $\delta_1\delta_2$ can be replaced by an auxiliary logical variable $\delta_3 = \delta_1\delta_2$, i.e., $[\delta_3 = 1] \leftrightarrow [\delta_1 = 1] \wedge [\delta_2 = 1]$, which is equivalent to the following linear constraints

$$\begin{cases} -\delta_1 + \delta_3 \leq 0, \\ -\delta_2 + \delta_3 \leq 0, \\ \delta_1 + \delta_2 - \delta_3 \leq 1. \end{cases} \tag{3.36}$$

Moreover, the product $\delta \tilde{f}(\tilde{x})$ can be replaced by the auxiliary real variable $z = \delta \tilde{f}(\tilde{x})$, which satisfies $[\delta = 0] \Rightarrow [z = 0]$ and $[\delta = 1] \Rightarrow [z = \tilde{f}(\tilde{x})]$. Then $z = \delta \tilde{f}(\tilde{x})$ is equivalent to

$$\begin{cases} z \leq \tilde{M}\delta, \\ z \geq \tilde{m}\delta, \\ z \leq \tilde{f}(\tilde{x}) - \tilde{m}(1-\delta), \\ z \geq \tilde{f}(\tilde{x}) - \tilde{M}(1-\delta). \end{cases} \tag{3.37}$$

It is noted that (3.35)–(3.37) yield linear inequalities since $\tilde{f}$ is affine.

3.4.2.2 The Mixed Logical Dynamic Model

In the interval $[s_k, s_{k+1}]$, the differential equation of the kinetic energy (3.14) can now be rewritten as

$$\frac{\mathrm{d}\tilde{E}}{\mathrm{d}s} = \frac{1}{m\rho}u(k) - \frac{2(a_2 + \xi_{2,k})}{\rho}\tilde{E}(s) - \frac{1}{\rho}(a_1 + \xi_{1,k}),$$

where $u(k)$ is a constant in the interval $[s_k, s_{k+1}]$. By defining $\zeta = \frac{1}{m\rho}$, $\eta_k = -\frac{2(a_2+\xi_{2,k})}{\rho}$, $\gamma_k = -\frac{1}{\rho}(a_1 + \xi_{1,k})$, this equation can be rewritten as

$$\frac{\mathrm{d}\tilde{E}}{\mathrm{d}s} = \zeta u(k) + \eta_k \tilde{E}(s) + \gamma_k. \tag{3.38}$$

We solve this differential equation with initial condition $\tilde{E}(s_k) = E(k)$ and we get the following formula for $E(s_k + 1)$:

$$E(s_k + 1) = e^{\eta_k \Delta s_k} E(s_k) + (e^{\eta_k \Delta s_k} - 1)\frac{\zeta}{\eta_k} u(k) + (e^{\eta_k \Delta s_k} - 1)\frac{\gamma_k}{\eta_k}, \qquad (3.39)$$

with $E(s_1) = \tilde{E}_{\text{start}}$ and $E(s_{N+1}) = \tilde{E}_{\text{end}}$.

Remark 3.4 For the sake of simplicity, we use $E(k)$ as a short-hand notation for $\tilde{E}(s_k)$ from now on.

Defining $a_k = e^{\eta_k \Delta s_k}$, $b_k = (e^{\eta_k \Delta s_k} - 1)\frac{\zeta}{\eta_k}$ and $c_k = (e^{\eta_k \Delta s_k} - 1)\frac{\gamma_k}{\eta_k}$, (3.39) can now be simplified as follows:

$$E(k + 1) = a_k E(k) + b_k u(k) + c_k. \qquad (3.40)$$

Note that this is an affine equation. For the differential equation (3.15), we approximate it as follows by employing a trapezoidal integration rule [35]:

$$t(k + 1) = t(k) + \frac{1}{2}\left(\frac{1}{\sqrt{2E(k)}} + \frac{1}{\sqrt{2E(k + 1)}}\right)\Delta s_k, \qquad (3.41)$$

with $t(1) = 0$. Furthermore, the nonlinear parts in (3.41) will be approximated by piecewise affine functions. There are various methods for approximating functions in a piecewise affine way, see, e.g., the overview by Azuma et al. [36]. In this chapter, we first select the number of regions of the piecewise affine function and then optimize the interval lengths and parameters of the affine functions using least-squares optimization to minimize the squared difference between the original function and the piecewise affine function. Recall that E_{min} denotes the minimum kinetic energy, so the maximum kinetic energy can be calculated by

$$E_{\text{max}} = \max_{k=1,2,\ldots,N} (E_{\text{max}}(k)) = \max_{k=1,2,\ldots,N} \left(\frac{1}{2} v_{\text{max}}^2(k)\right).$$

Hence, the nonlinear function $f(E) = \frac{1}{2\sqrt{2E}}$ should be approximated over the interval $[E_{\text{min}}, E_{\text{max}}]$ by a piecewise affine function. However, the speed limit depends on the space interval, i.e., different space intervals may have different speed limits, which may be less than the overall maximum of the speed limit. Therefore, we could adapt the coefficients of the piecewise affine approximations with respect to the space interval index k, i.e., we can have different piecewise affine subfunctions for different space intervals within valid speed intervals. In this way, the approximation error will be reduced. For example, if we consider an approximation using three affine subfunctions (cf. Fig. 3.2), the piecewise affine approximation[4] of the nonlinear function $f(E(k)) = \frac{1}{2\sqrt{2E(k)}}$ can be written as

[4]The approximation error can be reduced by taking more regions.

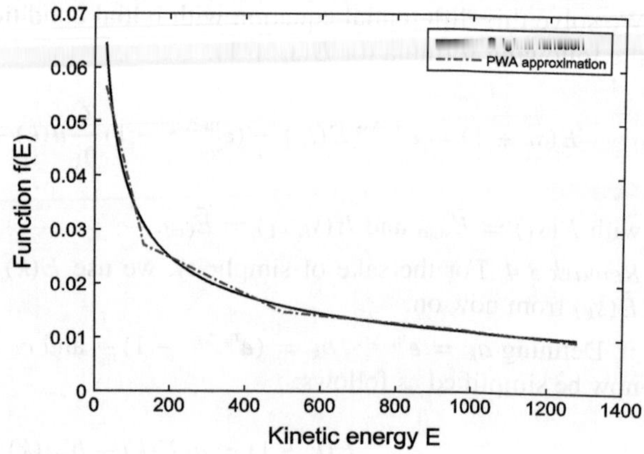

Fig. 3.2 The piecewise affine approximation of the nonlinear function $f(\cdot)$

$$f_{\text{PWA}}(E(k)) = \begin{cases} \alpha_{1,k}E(k) + \beta_{1,k} \text{ for } E_{0,k} \le E(k) \le E_{1,k}, \\ \alpha_{2,k}E(k) + \beta_{2,k} \text{ for } E_{1,k} \le E(k) \le E_{2,k}, \\ \alpha_{3,k}E(k) + \beta_{3,k} \text{ for } E_{2,k} \le E(k) \le E_{3,k}, \end{cases} \tag{3.42}$$

with $E_{0,k} = E_{\text{min}}$ and $E_{3,k} = E_{\text{max}}(k)$ for the interval $[s_k, s_{k+1}]$. The coefficients $\alpha_{1,k}$, $\beta_{1,k}$, $\alpha_{2,k}$, $\beta_{2,k}$, $\alpha_{3,k}$, $\beta_{3,k}$, and the values of $E_{1,k}$ and $E_{2,k}$ are determined by least-squares optimization.

Based on the piecewise affine approximations, the time dynamics (3.41) can then be approximated as

$$t(k+1) = t(k) + (\alpha_{l,k}E(k) + \beta_{l,k} + \alpha_{m,k+1}E(k+1) + \beta_{m,k+1})\Delta s_k, \tag{3.43}$$

with $E_{l-1,k} \le E(k) \le E_{l,k}$, $E_{m-1,k+1} \le E(k+1) \le E_{m,k+1}$ for $l, m \in \{1, 2, 3\}$.

Furthermore, the maximum traction force u_{max} is also a nonlinear function of the velocity as given in (3.3) or (3.4) that can be reformulated as a nonlinear function of the kinetic energy. In a similar way as the approximation of the nonlinear function $f(\cdot)$, we can obtain a piecewise affine approximation of the maximum traction force. If we consider an approximation[5] using three affine subfunctions (cf. (3.42)), then the approximation can be written as

$$u_{\text{max,PWA}}(E(k)) = \begin{cases} \lambda_{1,k}E(k) + \mu_{1,k} \text{ for } E_{4,k} \le E(k) \le E_{5,k}, \\ \lambda_{2,k}E(k) + \mu_{2,k} \text{ for } E_{5,k} \le E(k) \le E_{6,k}, \\ \lambda_{3,k}E(k) + \mu_{3,k} \text{ for } E_{6,k} \le E(k) \le E_{7,k}, \end{cases} \tag{3.44}$$

with $E_{4,k} = E_{\text{min}}$ and $E_{7,k} = E_{\text{max}}(k)$. Similarly, the coefficients of affine functions and the values of $E_{5,k}$ and $E_{6,k}$ are decided by least-squares optimization.

By introducing some auxiliary logical variables [37], the above piecewise affine model with piecewise affine constraints can be transformed into a mixed logical dynamic model. First, we consider piecewise affine approximation (3.42). In order

[5]For M affine subfunctions with $M > 3$ a similar procedure can be used.

to transform this equation, we introduce auxiliary logical variables $\delta_1(k)$ and $\delta_2(k)$, which are defined as

$$\begin{aligned}[E(k) \leq E_{1,k}] &\Leftrightarrow [\delta_1(k) = 1],\\ [E(k) \leq E_{2,k}] &\Leftrightarrow [\delta_2(k) = 1].\end{aligned} \tag{3.45}$$

Then we can rewrite (3.42) as follows:

$$\begin{aligned}f_{\text{PWA}}(E(k)) = &\delta_1(k)\delta_2(k)[\alpha_{1,k}E(k) + \beta_{1,k}] + (1 - \delta_1(k))\delta_2(k)[\alpha_{2,k}E(k) + \beta_{2,k}]\\ &+ (1 - \delta_1(k))(1 - \delta_2(k))[\alpha_{3,k}E(k) + \beta_{3,k}].\end{aligned} \tag{3.46}$$

Since the maximum and minimum values of $E(k)$ are $E_{\max}(k)$ and $E_{\min}$ for space interval s_k, s_{k+1}, based on the transformation property (3.35), the logical conditions (3.45) can be rewritten as linear inequalities

$$\begin{aligned}(E_{\max}(k) - E_{1,k})\delta_1(k) &\leq E_{\max}(k) - E(k),\\ (E_{\min} - E_{1,k} - \varepsilon)\delta_1(k) &\leq E(k) - E_{1,k} - \varepsilon,\\ (E_{\max}(k) - E_{2,k})\delta_2(k) &\leq E_{\max}(k) - E(k),\\ (E_{\min} - E_{2,k} - \varepsilon)\delta_2(k) &\leq E(k) - E_{2,k} - \varepsilon.\end{aligned} \tag{3.47}$$

Furthermore, an auxiliary logical variable $\delta_3(k)$ is introduced to replace the product $\delta_1(k)\delta_2(k)$. The condition $\delta_3(k) = \delta_1(k)\delta_2(k)$ can be rewritten as a system of linear inequalities according to (3.36)

$$\begin{aligned}-\delta_1(k) + \delta_3(k) &\leq 0,\\ -\delta_2(k) + \delta_3(k) &\leq 0,\\ \delta_1(k) + \delta_2(k) - \delta_3(k) &\leq 1.\end{aligned} \tag{3.48}$$

By defining new auxiliary variables $z_1(k) = \delta_1(k)E(k)$, $z_2(k) = \delta_2(k)E(k)$, and $z_3(k) = \delta_3(k)E(k)$, which can be expressed as a system of linear inequalities according to (3.37),

$$\begin{aligned}z_i(k) &\leq E_{\max}(k)\delta_i(k),\\ z_i(k) &\geq E_{\min}\delta_i(k),\\ z_i(k) &\leq E(k) - E_{\min}(1 - \delta_i(k)),\\ z_i(k) &\geq E(k) - E_{\max}(k)(1 - \delta_i(k)),\end{aligned} \tag{3.49}$$

for $i = 1, 2, 3$. Finally, the function $f_{\text{PWA}}(\cdot)$ can be rewritten as

$$\begin{aligned}f_{\text{PWA}}(E(k)) = &\begin{bmatrix} -\alpha_{3,k} & \alpha_{2,k} - \alpha_{3,k} & \alpha_{1,k} - \alpha_{2,k} + \alpha_{3,k}\end{bmatrix} \begin{bmatrix} z_1(k) & z_2(k) & z_3(k)\end{bmatrix}^T\\ &+ \begin{bmatrix}-\beta_{3,k} & \beta_{2,k} - \beta_{3,k} & \beta_{1,k} - \beta_{2,k} + \beta_{3,k}\end{bmatrix} \begin{bmatrix}\delta_1(k) & \delta_2(k) & \delta_3(k)\end{bmatrix}^T\\ &+ \alpha_{3,k}E(k) + \beta_{3,k}.\end{aligned} \tag{3.50}$$

In order to deal with the piecewise affine constraints of the maximum traction force (cf. (3.44)), auxiliary logical variables $\delta_4(k)$ and $\delta_5(k)$ are introduced, which are defined as

$$\begin{aligned}[E(k) \le E_{5,k}] &\Leftrightarrow [\delta_4(k) = 1],\\ [E(k) \le E_{6,k}] &\Leftrightarrow [\delta_5(k) = 1].\end{aligned} \tag{3.51}$$

Similar to (3.45), the logical conditions (3.51) can be recast as linear inequalities by applying transformation property (3.35). In addition, another binary variable $\delta_6(k)$ is introduced similarly as $\delta_3(k)$, and it is defined as $\delta_6(k) = \delta_4(k)\delta_5(k)$. Furthermore, auxiliary variables $z_4(k) = \delta_4(k)E(k)$, $z_5(k) = \delta_5(k)E(k)$, and $z_6(k) = \delta_6(k)E(k)$ are defined in order to rewrite the constraints into a system of linear inequalities. The piecewise affine constraints $u(k) \le u_{\text{max,PWA}}(E(k))$ can then be written as

$$\begin{aligned}u(k) \le& \left[-\lambda_{3,k}\ \lambda_{2,k} - \lambda_{3,k}\ \lambda_{1,k} - \lambda_{2,k} + \lambda_{3,k}\right]\left[z_4(k)\ z_5(k)\ z_6(k)\right]^T\\ &+ \left[-\mu_{3,k}\ \mu_{2,k} - \mu_{3,k}\ \mu_{1,k} - \mu_{2,k} + \mu_{3,k}\right]\left[\delta_4(k)\ \delta_5(k)\ \delta_6(k)\right]^T\\ &+ \lambda_{3,k}E(k) + \mu_{3,k}.\end{aligned} \tag{3.52}$$

Now the dynamics for the operation of a train can be rewritten as the following mixed logical dynamic model

$$x(k+1) = A_k x(k) + B_k u(k) + C_{1,k}\delta(k) + C_{2,k}\delta(k+1) + D_{1,k}z(k) + D_{2,k}z(k+1) + e_k, \tag{3.53}$$

where

$$x(k) = \left[E(k)\ t(k)\right]^T,\ A_k = \begin{bmatrix} a_k & 0\\ \Delta s_k(\alpha_{3,k} + a_k\alpha_{3,k+1}) & 1\end{bmatrix},\ B_k = \begin{bmatrix} b_k\\ \Delta s_k\alpha_{3,k+1}b_k\end{bmatrix},$$

$$C_{1,k} = \Delta s_k\begin{bmatrix} 0 & 0 & 0 & 0\ 0\ 0\\ -\beta_{3,k} & \beta_{2,k} - \beta_{3,k} & \beta_{1,k} - \beta_{2,k} + \beta_{3,k} & 0\ 0\ 0\end{bmatrix},$$

$$C_{2,k} = \Delta s_k\begin{bmatrix} 0 & 0 & 0 & 0\ 0\ 0\\ -\beta_{3,k+1} & \beta_{2,k+1} - \beta_{3,k+1} & \beta_{1,k+1} - \beta_{2,k+1} + \beta_{3,k+1} & 0\ 0\ 0\end{bmatrix},$$

$$D_{1,k} = \Delta s_k\begin{bmatrix} 0 & 0 & 0 & 0\ 0\ 0\\ -\alpha_{3,k} & \alpha_{2,k} - \alpha_{3,k} & \alpha_{1,k} - \alpha_{2,k} + \alpha_{3,k} & 0\ 0\ 0\end{bmatrix},$$

$$D_{2,k} = \Delta s_k\begin{bmatrix} 0 & 0 & 0 & 0\ 0\ 0\\ -\alpha_{3,k+1} & \alpha_{2,k+1} - \alpha_{3,k+1} & \alpha_{1,k+1} - \alpha_{2,k+1} + \alpha_{3,k+1} & 0\ 0\ 0\end{bmatrix},$$

and $e_k = \begin{bmatrix} c_k\\ \Delta s_k(\alpha_{3,k+1}c_k + \beta_{3,k} + \beta_{3,k+1})\end{bmatrix}.$

The mixed logical dynamic model (3.53) is subject to the linear constraints of the form (3.35)–(3.37) resulting from the transformation as well as the upper bound and lower bound constraints for $E(k)$, $t(k)$, and $u(k)$. All these constraints can be written more compactly as

$$R_{1,k}\delta(k) + R_{2,k}\delta(k+1) + R_{3,k}z(k) + R_{4,k}z(k+1) \leq R_{5,k}u(k) + R_{6,k}x(k) + R_{7,k}, \tag{3.54}$$

with appropriately defined coefficient matrices $R_{i,k}$, for $i = 1, 2, \ldots, 7$.

The objective function (3.16) can be discretized as

$$J = \sum_{k=1}^{N} u(k)\Delta s_k + \sum_{k=1}^{N-1} \lambda|\Delta u(k)|, \tag{3.55}$$

where $\Delta u(k) = u(k+1) - u(k)$. We introduce a new variable $\omega(k)$ to deal with the absolute value of $\Delta u(k)$, and we add the linear inequalities:

$$\begin{aligned} \omega(k) &\geq u(k+1) - u(k), \\ \omega(k) &\geq u(k) - u(k+1). \end{aligned} \tag{3.56}$$

Since $\lambda > 0$, minimizing (3.55) is equivalent to minimizing

$$\tilde{J} = \sum_{k=1}^{N} u(k)\Delta s_k + \sum_{k=1}^{N-1} \lambda\omega(k), \tag{3.57}$$

subject to (3.56). Indeed, it is easy to verify that when we minimize the objective function (3.57) subject to (3.56), the optimal value of $\omega(k)$ will be equal to $|\Delta u(k)|$, so (3.55) will also be minimized.

3.4.2.3 The Mixed Integer Linear Programming problem

The optimal control problem can then be recast as an MILP problem, where some of decision variables are binary variables (i.e., $\tilde{\delta}$) and some are real variables (i.e., $\tilde{u}, \tilde{\omega}, \tilde{z}$) with

$$\tilde{u} = \begin{bmatrix} u(1) \\ u(2) \\ \vdots \\ u(N) \end{bmatrix}, \quad \tilde{\delta} = \begin{bmatrix} \delta(1) \\ \delta(2) \\ \vdots \\ \delta(N+1) \end{bmatrix}, \quad \tilde{z} = \begin{bmatrix} z(1) \\ z(2) \\ \vdots \\ z(N+1) \end{bmatrix}, \quad \tilde{\omega} = \begin{bmatrix} \omega(1) \\ \omega(2) \\ \vdots \\ \omega(N-1) \end{bmatrix}.$$

Furthermore, if we define $\tilde{V} = \begin{bmatrix} \tilde{u}^T \ \tilde{\delta}^T \ \tilde{z}^T \ \tilde{\omega}^T \end{bmatrix}^T$, the equivalent formulation of the optimal control problem is obtained as follows:

$$\min_{\tilde{V}} C_J^T \tilde{V} \tag{3.58}$$

subject to

$$F_1 \tilde{V} \leq F_2 x(1) + f_3, \tag{3.59}$$

$$F_4 \tilde{V} = F_5 x(1) + f_6, \tag{3.60}$$

where $C_J = \left[\Delta s_1 \cdots \Delta s_N \; 0 \cdots 0 \; \lambda \cdots \lambda \right]^T$. This can be shown as follows. The constraints for the MILP problem (3.54) are considered for $k = 1, 2, \ldots, N$. We can substitute $x(k)$ in the constraints using the state equation (3.53) recursively. The substituted form is obtained as the following expression:

$$\begin{aligned} x(k) = &\Big[\prod_{j=1}^{k-1} A_j\Big] x(1) + \sum_{i=1}^{k-1}\Big[\prod_{j=i+1}^{k-1} A_j\Big] B_i u(i) + \Big[\prod_{j=2}^{k-1} A_j\Big] C_{1,1}\delta(1) \\ &+ \sum_{i=2}^{k-1}\Big[\prod_{j=i+1}^{k-1} A_j\Big](A_i C_{i-1,2} + C_{i,1})\delta(i) + C_{k-1,2}\delta(k) \\ &+ \Big[\prod_{j=2}^{k-1} A_j\Big] D_{1,1} z(1) + \sum_{i=2}^{k-1}\Big[\prod_{j=i+1}^{k-1} A_j\Big](A_i D_{i-1,2} + D_{i,1}) z(i) \\ &+ D_{k-1,2} z(k) + \sum_{i=1}^{k-1}\Big[\prod_{j=i+1}^{k-1} A_j\Big] e_i. \end{aligned}$$

In addition, the end point condition $x(N+1) = [E_{\text{end}} \; T]^T$ needs to be considered in (3.60). Because we know the value of $x(N+1)$, the values of α_m and β_m in (3.43) are also known. So the state equation at the end point can be written as

$$x(N+1) = A_N x(N) + B_N u(N) + C_{1,N}\delta(N) + D_{1,N} z(N) + e_N,$$

where $A_N = \begin{bmatrix} a_N & 0 \\ \Delta s_N(\alpha_{3,N} + \alpha_{m,N+1} a_N) & 1 \end{bmatrix}$, $B_N = \begin{bmatrix} b_N \\ \Delta s_N \alpha_{m,N+1} b_N \end{bmatrix}$, and $e_N = \begin{bmatrix} c_N \\ \Delta s_N(\alpha_{m,N+1} c_N + \beta_{m,N+1} + \beta_{3,N}) \end{bmatrix}$. By properly defining F_1, F_2, f_3, F_4, F_5, and f_6, we can write all these constraints in the form (3.59) and (3.60). The MILP problem (3.58)–(3.60) can be solved by several existing commercial and free solvers, such as CPLEX, Xpress-MP, GLPK (see, e.g., [38, 39]).

Remark 3.5 The variable $t(s_{k_j})$ in (3.21) and (3.22) is one of the state variables of the model (3.53) and (3.54). Note that (3.21) and (3.22) are linear constraints. Hence, we still have an MILP problem if the target point and target window constrains are included. See [3] for more information.

3.5 Case Study

In order to demonstrate the performance of the pseudospectral method and the MILP approach, we use a case study (inspired by Vašak et al. [6]) to compare these two approaches to the DDP approach proposed in [40]. The reason for selecting the DDP approach for comparison is that [40] concludes that the performance of the DDP approach is better than that of the sequential quadratic programming approach and the coasting strategy obtained by the maximum principle. The optimal trajectories obtained by those the pseudospectral, MILP approach, and DDP approaches are compared with each other. In addition, both the computation time and the performance with respect to the optimization objective and constraints violations of those approaches are analyzed.

3.5.1 Set-Up

The case study in this chapter is inspired by that of [6], where the track length between the departure station and arrival station is 10 km. In [6], there were no speed limit and grade profile. We add them as shown in Fig. 3.3. The rolling stock includes an SBB Re 460 locomotive [40–42], the parameters of which are presented in Table 3.1. The rotating mass factor is often chosen as 1.06 in the literature [9] and therefore we also adopt this value. Based on the assumption made in Sect. 3.3, the unit kinetic energy should be larger than zero. In this test case, the minimum kinetic energy is chosen as 0.1 J. The maximum traction force of the SBB Re 460 locomotive is a nonlinear function of the train's velocity and the maximum value of

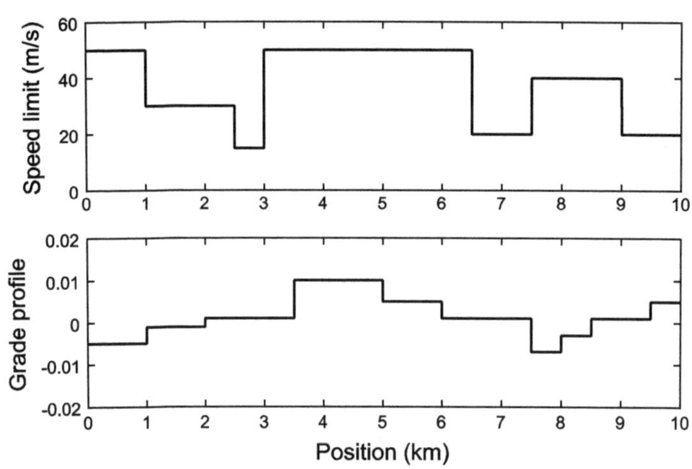

Fig. 3.3 The speed limits and the grade profile along the track

Table 3.1 Parameters of the train and the line path

Property	Symbol	Value
Train mass (kg)	m	5.070×10^5
Basic resistance (N/kg)	R_b	$1.400 \times 10^{-2} + 2.564 \times 10^{-5}v^2$
Mass factor (–)	ρ	1.060
Line length (m)	s_T	1.000×10^4
Minimum kinetic energy (J)	$E_{\min}$	1.000×10^{-1}
Maximum braking force (regular) (N)	$u_{\max}^{\mathrm{dec}}$	4.475×10^5

this function is 300 kN as shown in Fig. 3.1. The objective function of the optimal train control problem considered here is a weighted sum of the energy consumption and passenger comfort, where the weight λ is taken as 500. The total running time for this trip is given by the timetable or the rescheduling process. Here, the total running time is 450 s, which consists of the minimum running time and 5 % running time supplements.

Two cases will be considered here:

- Case A: The maximum traction force is constant.
- Case B: The maximum traction force is a nonlinear function of the velocity.

In Case A, like the case study in [6], we assume that the maximum traction force $E_{\max}$ is constant: $E_{\max} = 300$ kN. First, the optimal trajectory planning problem is solved using PSOPT [31], which implements a pseudospectral method. In this case study, the problem is solved using the Legendre pseudospectral discretizations, with local automatic mesh refinement, starting with 40 nodes.

Second, the problem is solved using the MILP approach. Since the maximum traction force is constant, in this case the linear constraints caused by the piecewise affine approaximation (3.44) of the maximum traction force will not be taken into account here. In this chapter, the piecewise affine approximations of the nonlinear function $f(E) = \frac{1}{2\sqrt{2E}}$ depend on the space interval index k as stated in Sect. 3.4.2, i.e., we can have different piecewise affine subfunctions for different space intervals. In Fig. 3.3 there are five speed limits, i.e., 15, 20, 30, 40, and 50 m/s. Hence, five different approximations with two subfunctions of $f(\cdot)$ are obtained, the parameters of which are given in Table 3.2. For each space interval, an appropriate piecewise affine approximation can be chosen based on the given speed limit. In addition, we introduce two additional piecewise affine approximations of $f(\cdot)$ for the first space interval $[s_{\mathrm{start}}, s_2]$ and the last space interval $[s_N, s_{\mathrm{end}}]$. The train speed in these intervals are usually low; hence, in order to obtain a good fit we apply a weighted least-squares optimization to optimize the coefficients of the piecewise affine approximations for these intervals, where the weight function should have a high value in the low-speed range. The parameters of the piecewise affine approximations for the first and the last space intervals are presented in Table 3.3.

The length Δs_k for the interval $[s_k, s_{k+1}]$ depends on the speed limits, gradient profile, tunnels, and so on. If the number of space intervals N is larger, then the

Table 3.2 The piecewise affine approximations of the nonlinear function $f(\cdot)$

Approx. no.	Segment m	α_m ((s/m)^3)	β_m (s/m)	$E_m - E_{m+1}$ ((m/s)^2)
Approx. 1	Segment 1	-5.094×10^{-4}	0.077	0.1 – 71.2
	Segment 2	-1.739×10^{-4}	0.053	71.2 – 112.5
Approx. 2	Segment 1	-3.115×10^{-4}	0.067	0.1 – 115
	Segment 2	-6.719×10^{-5}	0.038	115 – 200
Approx. 3	Segment 1	-9.498×10^{-5}	0.044	0.1 – 240
	Segment 2	-2.347×10^{-5}	0.027	240 – 450
Approx. 4	Segment 1	-4.424×10^{-5}	0.035	0.1 – 415
	Segment 2	-9.646×10^{-6}	0.020	415 – 800
Approx. 5	Segment 1	-1.812×10^{-5}	0.025	0.1 – 640
	Segment 2	-6.213×10^{-6}	0.018	640 – 1250

Table 3.3 The piecewise affine approximations of the nonlinear function $f(\cdot)$ for the first and the last space intervals

Approx. no.	Segment m	α_m ((s/m)^3)	β_m (s/m)	$E_m - E_{m+1}$ ((m/s)^2)
Approx. 6	Segment 1	-4.645×10^{-4}	0.073	0.1 – 80.8
	Segment 2	-7.195×10^{-4}	0.042	80.8 – 312.5
Approx. 7	Segment 1	-1.446×10^{-4}	0.053	0.1 – 229.9
	Segment 2	-1.451×10^{-6}	0.023	229.9 – 450

computation time of the MILP approach will be longer, but the accuracy will be better. According to the speed limits and gradient given in Fig. 3.3, the length of each interval is chosen to equal 500 m, i.e., $\Delta s_k = 500$ m for $k = 1, 2, \ldots, 20$, which provides a good trade-off between the computation time and the accuracy. As MILP solver, we use CPLEX, implemented through the cplex interface function of the MATLAB Tomlab toolbox. For the DDP approach, the continuous nonlinear model of train (3.14) and (3.15) is discretized in space. The number of the space intervals is 100 and the length of each space interval is 100 m. To compute the optimal trajectory with DDP, we use a MATLAB function for dynamic programming that was introduced in [43].

For Case B, we consider a varying maximum traction force as shown in Fig. 3.1, the coefficients of which according to (3.3) are based on [40–42] and listed here in Table 3.4. In PSOPT, non-smooth path constraints can be handled by introducing a smooth version of the Heaviside function (see Sect. 3.4.1.2). In the MILP approach,

Table 3.4 The coefficients of the varying maximum traction force

Segment j	$c_{1,j}$ (kg · m/s^2)	$c_{2,j}$ (kg/s)	$c_{3,j}$ (kg/m)	$v_j - v_{j+1}$ (m/s)
1	3.000×10^5	-1.130×10^3	0	0 – 22.22
2	7.236×10^5	-2.726×10^4	3.128×10^2	22.22 – 38.89
3	4.237×10^5	-1.120×10^4	1.000×10^2	38.89 – 50

Table 3.5 The coefficients of the piecewise affine approximation of maximum traction force

Segment m	λ_m (kg/m)	μ_m (kg· m/s^2)	$E_m - E_{m+1}$ ((m/s)^2)
1	-1.283×10^2	2.974×10^5	0.1 – 250
2	-3.739×10^2	3.588×10^5	250 – 550
3	-0.636×10^2	1.881×10^5	550 – 1250

we need to approximate the nonlinear maximum traction force by piecewise affine functions in (3.44), where the coefficients may also depend on the space interval index k. Here, for simplicity, we use only one piecewise affine approximation with three affine subfunctions for all k. The parameters of the piecewise affine function are listed in Table 3.5.

3.5.2 Results and Discussion

3.5.2.1 Results for Case A: The Maximum Traction Force Is Constant

The optimal solution of the pseudospectral method using PSOPT, which is obtained after 7 mesh refinement iterations, has 179 nodes. The calculation time for PSOPT is 370 s on a 1.8 GHz Intel Core2 Duo CPU running a 64-bit Linux operating system and the computation time for DDP is 2 min and 8 s with 100 space intervals as shown in Table 3.6. However, the calculation time for the MILP approach is 0.34 s on the same CPU and operation system as above, which is much shorter than the calculation time of PSOPT and DDP.

Figure 3.4 shows the optimal control inputs with constant maximum traction force, where the solid line, the dash-dotted line, and the dashed line represent the results calculated by MILP, PSOPT, and DDP, respectively. It can be seen from Fig. 3.4 that the results obtained by these three approaches show a similar trend, but there exist more discrete changes but with a smaller magnitude in the control inputs of PSOPT and DDP. This is mainly because of the larger number of space intervals. More specifically, there are 178 and 100 space intervals in PSOPT and DDP, respectively.

Table 3.6 Performance comparison of PSOPT, MILP, and DDP for Case A with constant maximum traction force

Performance index	PSOPT	MILP	DDP
$J_{\min}$ (–)	2.424×10^8	2.695×10^8	2.473×10^8
CPU time (s)	370	0.34	127
End position s_{end} violation (m)	0	0	0
End kinetic energy E_{end} violation (m/s^2)	0.1	0.172	0.576
End time T_{end} violation (s)	0.496	9.142	5.093
Speed limit violation (–)	No	No	No

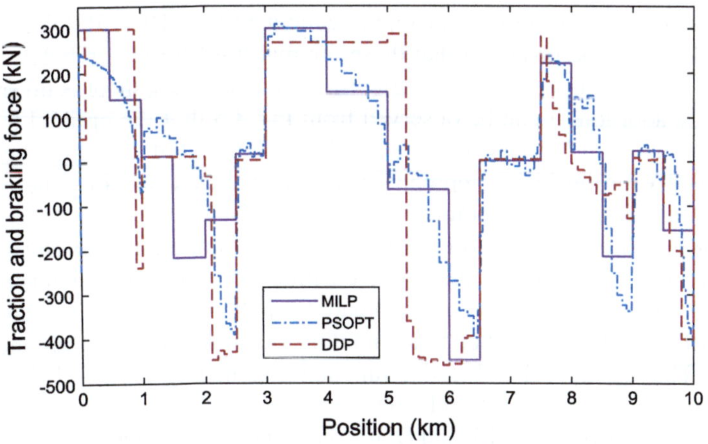

Fig. 3.4 The optimal control inputs with constant maximum traction force for the following approaches: MILP, PSOPT, and DDP

However, there are just 20 space intervals in the MILP approach. The optimal control inputs calculated by these three approaches are applied to the nonlinear continuous-time train model (3.1)–(3.2). The differential equation of the nonlinear model is solved numerically using a variable step Runge–Kutta method and the resulting speed–position trajectories for the train are shown in Fig. 3.5. The dotted line shows the speed limit for the trip, which is determined by the characteristics of the train, line, etc. The solid line, dash-dotted line, and dashed line represent the optimal

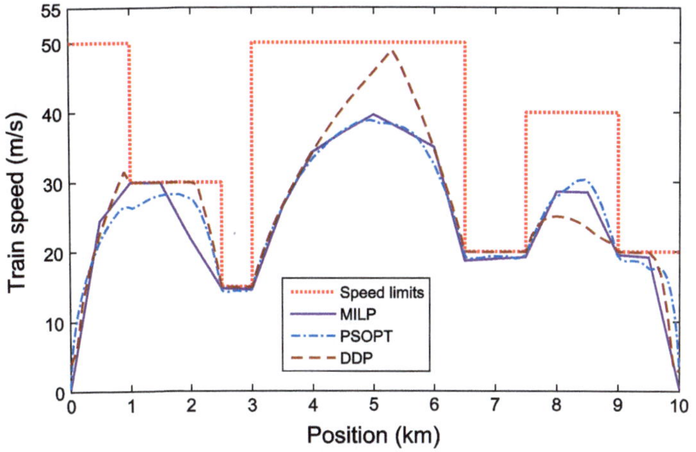

Fig. 3.5 The trajectories generated by the nonlinear continuous-time train model (3.1) and (3.2) using optimal control inputs with constant maximum traction force for the following approaches: MILP, PSOPT, and DDP

trajectories obtained using control inputs generated by MILP, PSOPT, and DDP, respectively. It can be observed that these optimal trajectories are below the speed limit. This means that the speed constraints are satisfied, i.e., there is no speed limit violation. In addition, it can be observed from Fig. 3.5 that the optimal trajectories obtained using control inputs generated by PSOPT and DDP are smoother than the one obtained with the MILP approach, which is mainly caused by the number of space intervals as mentioned before.

In Table 3.6, the values of the objective function, computation time, and constraints violations (i.e., speed limit violation, end position violation, end kinetic energy violation, and end time violation) are compared for the control inputs generated by PSOPT, MILP, and DDP applied to the nonlinear continuous-time train model (3.1) and (3.2). The values of the objective function obtained by the PSOPT, MILP, and DDP approach are 2.424×10^8 and 2.695×10^8, and 2.473×10^8, respectively. The relative differences of the MILP and DDP control performance are 11.18 and 2.02 % of that of the pseudospectral method. Therefore, the pseudospectral approach yields the smallest objective value and the constraints violations for the pseudospectral method are also the smallest.

3.5.2.2 Results for Case B: The Maximum Traction Force Is a Nonlinear Function

Figure 3.6 shows the optimal control inputs for Case B, where the solid line, the dash-dotted line, and the dashed line represent the optimal control inputs obtained using MILP, PSOPT, and DDP, respectively. When we compare Figs. 3.4, 3.5, and 3.6, the maximum traction force in Fig. 3.6 is no longer equal to 300 kN for the MILP

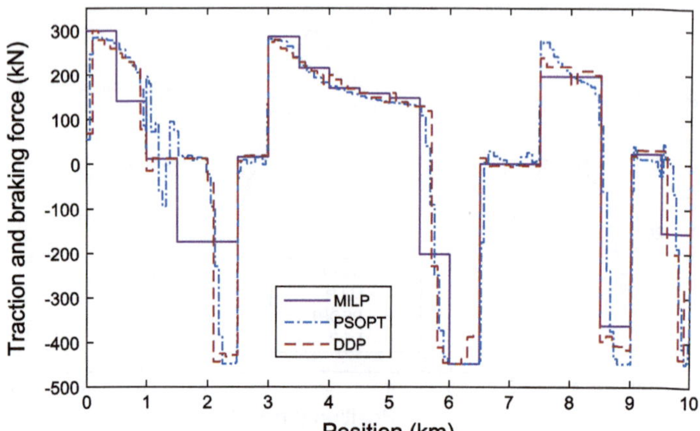

Fig. 3.6 The optimal control inputs with varying maximum traction force for the following approaches: MILP, PSOPT, and DDP

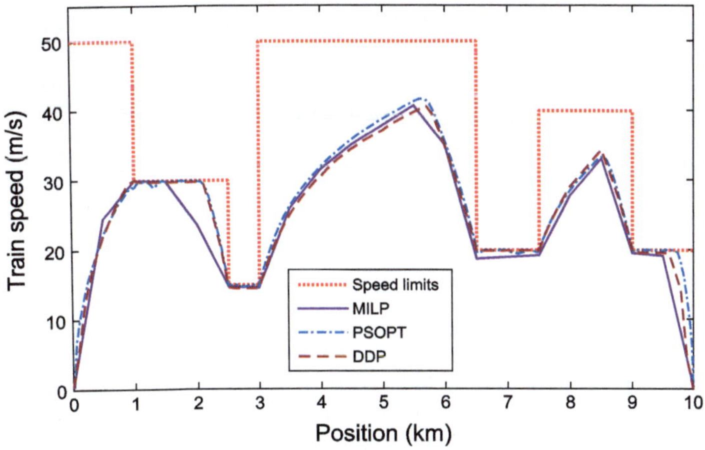

Fig. 3.7 The trajectories generated by the nonlinear continuous-time train model (3.1) and (3.2) using optimal control inputs with varying maximum traction force for the following approaches: MILP, PSOPT, and DDP

approach in the space interval [3000, 4000], but it becomes smaller and smaller when the speed increases. This is because the varying maximum traction force is decreasing when the speed goes up. Similar results can be observed for the optimal inputs obtained by PSOPT and DDP. Figure 3.7 shows the speed–position trajectories for the train under varying maximum traction force constraints when applying these inputs to the nonlinear train model (3.1) and (3.2). The dotted line, solid line, dash-dotted line, and dashed line show the speed limits and the trajectories obtained by control inputs of the MILP, PSOPT, and DDP, respectively. All these trajectories are below the speed limits, so there is no speed limit violation. Furthermore, we can see from Fig. 3.7 that the slopes of these three optimal trajectories obtained in the space interval [3500, 5000] are smaller than those of Fig. 3.5. The reason is that the maximum traction force decreases with the increase of the train's speed.

The values of the objective function, the computation time, and the constraints violations are compared for PSOPT, MILP, and DDP in Table 3.7. Similar as the results in Case A, the pseudospectral approach obtains the minimum objective function value 2.625×10^8. Note that it is higher than that in Table 3.6 because of the consideration of the varying maximum traction force. The relative differences of the MILP and DDP approach in control performance are 7.31 and 2.17 % when compared to that of the pseudospectral method. PSOPT includes a local automatic mesh refinement. In this case study, we take 40 nodes as initial value. The final solution is obtained after 7 mesh refinement iterations and has 199 nodes. The calculation time is 1147 s. Compared with the problem in Case A, 20 nodes are added and the computation time is almost 13 min longer. For the DDP approach, the computation time is 135 s, which is 8 s longer than that of Case A. In the MILP approach, for each space interval an extra binary variable and an auxiliary real variable are introduced

Table 3.7 Performance comparison of PSOPT, MILP, and DDP for Case B with varying maximum traction force

Performance index	PSOPT	MILP	DDP
J_{min} (–)	2.625×10^8	2.817×10^8	2.682×10^8
CPU time (s)	1147	0.52	135
End position s_{end} violation (m)	0	0	0
End kinetic energy E_{end} violation (m/s^2)	0.1	0.104	0.083
End time T_{end} violation (s)	0	4.072	5.324
Speed limit violation (–)	No	No	No

in the mixed logical dynamic model compared with Case A, since the maximum traction force is considered as a nonlinear function that is approximated by a piecewise affine approximation with 2 subfunctions. Therefore, 40 variables are added to the MILP problem since the number of the space intervals is 20. The computation time is now 0.52 s, which is larger than the 0.34 s of Case A, but it still is much lower than the computation time the pseudospectral method and the DDP approach. Similar to the results shown in Table 3.6, there are no speed limit violations and the end kinetic energy violation is very small. Furthermore, the end time violation for the pseudospectral method is also very small, but for the MILP and DDP approach this violation is about 1 % of the total running time.

3.5.2.3 Discussion

It is concluded that for the given case study the pseudospectral approach obtains the best control performance, which considers the value of the objective function and the constraints violations. However, when the computation time is also taken into consideration, the MILP approach yields the best overall performance.

It is worth noting that we apply a trapezoidal integration rule to approximate the time differential equation (3.15) and then use piecewise affine functions to approximate the nonlinear function (3.41) in the MILP approach. Therefore, the end time violation in Table 3.6 of the MILP approach is probably caused by these approximations. The approximation error can be reduced further by adjusting the piecewise affine approximations according to Footnote 4 on p. 35, but then the computation time would go up.

3.6 Summary

In this chapter, we have considered the optimal trajectory planning problem for a single train, where two approaches, i.e., the pseudospectral method and the MILP approach, have been presented to solve this problem. In the pseudospectral method, the optimal trajectory planning problem is formulated as a multiple-phase optimal control problem based on piecewise line resistance and speed limits. The constraints caused by the varying maximum traction force are defined as nonlinear path constraints. In the MILP approach, the nonlinear train operation model is formulated as a mixed logical dynamic model via piecewise affine approximations. The optimal control problem is then recast as an MILP problem. The case study shows that the pseudospectral method has the best control performance and the MILP has the best overall performance if the computation time is also taken into account. Furthermore, the computation time of the MILP approach is much shorter compared with that of the pseudospectral method and the DDP approach. For the given case study, the relative difference between the performance of the MILP approach and that of the pseudospectral approach is about 10 %.

When the timetable is known, the two proposed approaches (i.e., the MILP and the pseudospectral approach) can be used to calculate the optimal trajectory for trains between two consecutive stations to save energy and to ensure passenger comfort. If there are some disturbances in the network, then one could use a rescheduling approach to reorder trains and determine new timetables [9, 14]. Next, the affected trains have to optimize their trajectories according to the new train schedule. In this case, the trajectory planning problem needs to be solved quickly to satisfy the real-time requirements; so then the MILP approach could be applied since it gives the best trade-off between computational speed and accuracy.

An extensive comparison and assessment of the pseudospectral method, the MILP approach, and other approaches in the literature for various case studies and a wide range of scenarios will be a topic for future work. In addition, in this chapter we have focused on the trajectory planning for a single train between two stations with the assumption that the constraints and disturbances caused by signaling systems and other trains are handled by the lower control level. However, in practice these constraints and disturbances are significant for the capacity of the railway network, and therefore some interesting conflict detection and resolution approaches have been proposed to manage these constraints and disturbances during the rescheduling phases [44–46]. In future work, one could combine these conflict detection and resolution approaches with the trajectory planning approaches proposed here to solve the trajectory planning for multiple trains. There one could also use the MILP approach (including hierarchical and distributed optimization if the problem grows too large). Furthermore, the pseudospectral and MILP solvers used in this chapter are general-purpose solvers. By making use of the specific structure and properties of the optimal trajectory planning problem, significant speed-ups can be expected. Therefore, tailored pseudospectral and MILP solvers for the optimal trajectory planning problem for trains could be developed.

References

1. Wang Y, De Schutter B, Ning B, Groot N, van den Boom T (2011) Optimal trajectory planning for trains using mixed integer linear programming. In: Proceedings of the 14th international IEEE conference on intelligent transportation systems (ITSC 2011). Washington DC, USA, pp 1598–1604
2. Wang Y, De Schutter B, van den Boom T, Ning B (2013b) Optimal trajectory planning for trains—a pseudospectral method and a mixed integer linear programming approach. Transp Res Part C 29:97–114
3. Wang Y, De Schutter B, van den Boom T, Ning B (2012) Optimal trajectory planning for trains under operational constraints using mixed integer linear programming. In: Proceedings of the 13th IFAC symposium on control in transportation systems (CTS'2012). Sofia, Bulgaria, pp 158–163
4. Elnagar G, Kazemi M, Razzaghi M (1995) The pseudospectral Legendre method for discretizing optimal control problems. IEEE Trans Autom Control 40:1793–1796
5. Gong Q, Kang W, Bedrossian N, Fahroo F, Sekhavat P, Bollino K (2007) Pseudospectral optimal control for military and industrial applications. In: Proceedings of the 46th IEEE conference on decision and control. New Orleans, LA, USA, pp 4128–4142
6. Vašak M, Baotić M, Perić N, Bago M (2009) Optimal rail route energy management under constraints and fixed arrival time. In: Proceedings of the european control conference. Budapest, Hungary, pp 2972–2977
7. Franke R, Terwiesch P, Meyer M (2003) An algorithm for the optimal control of the driving of trains. In: Proceedings of the 39th IEEE conference on decision and control. Australia, Sydney, pp 2123–2128
8. Liu R, Golovicher I (2003) Energy-efficient operation of rail vehicles. Transp Res Part A: Policy Practice 37:917–931
9. Hansen I, Pachl J (2008) Railway, Timetable and traffic: analysis, modelling, simulation. Eurailpress, Hamburg, Germany
10. Rochard B, Schmid F (2000) A review of methods to measure and calculate train resistances. In: Proceedings of the Institution of Mechanical Engineers, Part F: Journal of Rail and Rapid Transit, vol. 214, pp 185–199
11. Mao B (2008) The calculation and design of train operations. People Transport press, Beijing
12. Huerlimann D, Nash A (2003) OPENTRACK—Simulation of railway networks, User Manual Version 1.3. Institute for Transportation Planning and Systems, ETH Zürich, Switzerland
13. Gao R (2008) Railway signal operation basis. China Railway Publishing House, Beijing
14. D'Ariano A, Pacciarelli D, Pranzo M (2008) Assessment of flexible timetables in real-time traffic management of a railway bottleneck. Transp Res Part C: Emerg Technol 16:232–245
15. Krasemann JT (2012) Design of an effective algorithm for fast response to the re-scheduling of railway traffic during disturbances. Transp Res Part C: Emerg Technol 20:62–78
16. Chang C, Xu D (2000) Differential evolution based tuning of fuzzy automatic train operation for mass rapid transit system. IEE Proc—Electric Power Appl 147:206–212
17. Rao X, Montigel M, Weidmann U (2013) Methods to improve railway capacity by integration of automatic train operation with centralized traffic management. In: Proceedings of the 5th international seminar on railway operations modelling and analysis (RailCopenhagen), Copenhagen, Denmark
18. Khmelnitsky E (2000) On an optimal control problem of train operation. IEEE Trans Autom Control 45:1257–1266
19. Howlett P (2000) The optimal control of a train. Ann Oper Res 98:65–87
20. Albrecht T, Gassel C, Binder A, van Luipen J (2010) Dealing with operational constraints in energy efficient driving. In: IET conference on railway traction systems (RTS 2010). Birmingham, UK, pp 1–7
21. Albrecht T, Gassel C, Binder A (2011) An overview on real-time speed control in rail-bound public transportation systems. In: Proceedings of the 2nd international conference on models and technologies for intelligent transportation systems. Belgium, Leuven, pp 1–4

22. Canuto C, Hussaini M, Quarteroni A, Zang T (1988) Spectral methods in fluid dynamics. Springer, New York
23. Ross I, Fahroo F (2003) Legendre pseudospectral approximations of optimal control problems. In: Kang W, Borges C, Xiao M (eds) New trends in nonlinear dynamics and control and their applications, volume 295 of Lecture Notes in Control and Information Science. Springer, Berlin, Germany, pp. 327–342
24. Gong Q, Ross I, Kang W, Fahroo F (2008) Connections between the covector mapping theorem and convergence of pseudospectral methods for optimal control. Comput Optim Appl 41:307–335
25. Fornberg B, Sloan D (1994) A review of pseudospectral methods for solving partial differential equations. Acta Numerica 3:203–267
26. Ross I, Fahroo F (2004) Pseudospectral knotting methods for solving optimal control problems. J. Guidance Control Dynam 27:397–405
27. Rutquist P, Edvall M (2008) PROPT: MATLAB optimal control software. Tomlab Optimization Inc, Pullman
28. Ross I (2004) User's manual for DIDO: a MATLAB application package for solving optimal control problems. Tomlab Optimization Inc, Pullman
29. Rao A, Benson D, Darby C, Patterson M, Francolin C, Sanders I (2010) Algorithm 902: Gpops, a matlab software for solving multiple- phase optimal control problems using the Gauss pseudospectral method. ACM Trans Math Softw 37:22:1–22:39
30. Becerra V (2010) Solving complex optimal control problems at no cost with psopt. In: Proceedings of IEEE multi-conference on systems and control. Yokohama, Japan, pp 1391–1396
31. Becerra VM (2010) Psopt optimal control solver user mannual - release 3
32. Kanwal R (1983) Generalized functions: theory and technique. Academic Press, New York
33. Betts J (1998) Survey of numerical methods for trajectory optimization. J Guidance Control Dynam 21:193–207
34. Williams H (1999) Model building in mathematical programming, 4th edn. Wiley, Chichester
35. Atkinson K (1978) An introduction to numerical analysis. Wiley, New York
36. Azuma S, Imura J, Sugie T (2010) Lebesgue piecewise affine approximation of nonlinear systems. Nonlinear Anal: Hybrid Syst 4:92–102
37. Bemporad A, Morari M (1999) Control of systems integrating logic, dynamics, and constraints. Automatica 35:407–427
38. Linderoth J, Ralphs T (2005) Noncommercial software for mixed-integer linear programming. In: Karlof J (ed) Integer programming: theory and practice, Operations Research Series. CRC Press, Boca Raton, FL, USA, pp 253–303
39. Atamturk A, Savelsbergh M (2005) Integer-programming software systems. Ann Oper Res 140:67–124
40. Franke R, Meyer M, Terwiesch P (2002) Optimal control of the driving of trains. Automatisierungstechnik 50:606–614
41. Gerber P (2001) Class Re465 locomotives for heavy-haul freight service. In: Proceedings of the Institution of Mechanical Engineers Part F: Journal of Rail and Rapid Transit vol. 215, pp 25–35
42. Schank T (2011) A fast algorithm for computing the running-time of trains by infinitesimal calculus. In: Proceedings of RAILROME 2011. Italy, Rome, pp 1–16
43. Sundström O, Guzzella L (2009) A generic dynamic programming matlab function. In: Proceedings of the 18th IEEE international conference on control applications, Part of 2009 IEEE multi-conference on systems and control. Saint Petersburg, Russia, pp 1625–1630
44. D'Ariano A, Albrecht T (2006) Running time re-optimization during real-time timetable perturbations. Computers in Railways X. WIT Press, Prague, Czech Republic, pp 531–540
45. D'Ariano A, Pranzoand M, Hansen I (2007) Conflict resolution and train speed coordination for solving real-time timetable perturbations. IEEE Trans Intell Transp Syst 8:208–222
46. Corman F, D'Ariano A, DPacciarelli P, Pranzo M, (2009) Evaluation of green wave policy in real-time railway traffic management. Transp Res Part C 17:607–616

Chapter 4
Optimal Trajectory Planning for Multiple Trains

Abstract In this chapter, the optimal trajectory planning problem for multiple trains under fixed block signaling systems and moving block signaling systems is considered. Two solution approaches are proposed to solve this optimal control problem for multiple trains: the greedy approach and the simultaneous approach. The greedy approach optimizes the trajectories of trains sequentially, where first the trajectory of the leading train is optimized and then the trajectory of the following train is optimized based on the trajectory of the leading train. In the simultaneous approach, the trajectories of all the trains are optimized at the same time. In each approach, the trajectory planning problem is similar to the problem of Chap. 3, and therefore it can also be solved using the pseudospectral method and the mixed integer linear programming (MILP) approach. The performance of the proposed approaches is compared via a case study. This chapter is based on Wang et al. (Control Eng Pract 22:44–56, 2014) [1] and is supported by the results presented in Wang et al. (Proceedings of the 12th European control conference, Zürich, Switzerland, pp 4556–4561, 2013) [2]; Wang et al. (Proceedings of the 5th international seminar on railway operations modelling and analysis (RailCopenhagen), Copenhagen, Denmark, 2013) [3].

4.1 Introduction

The approaches proposed in Chap. 3 ignore the impact caused by signaling systems, e.g., a fixed block signaling (FBS) system or a moving block signaling (MBS) system (see Sect. 2.1.2 for more detailed information about signaling systems). In this chapter, the constraints caused by the leading train in an FBS system and an MBS system are included into the trajectory planning problem. For the mixed integer linear programming (MILP) approach, the constraints caused by signaling systems are discretized and then recast as linear constraints by piecewise affine approximations. Hence, these constraints can easily be included into the MILP formulation [1–3].

The remainder of this chapter is structured as follows. In Sect. 4.2, the train model is summarized based on Sect. 3.2.1. In addition, the constraints for the operation of trains in an FBS system and an MBS system are also formulated in this section. Section 4.3 gives the mathematical formulation of the trajectory planning problem for

Y. Wang et al., *Optimal Trajectory Planning and Train Scheduling for Urban Rail Transit Systems*, Advances in Industrial Control,
DOI 10.1007/978-3-319-30889-0_4

multiple trains. In Sect. 4.4, we propose two solution schemes to solve the trajectory planning problem for multiple trains. For the MILP approach, Sect. 4.5 presents the transformation process of the FBS constraints and the MBS constraints. In addition, the mode vector constraints are proposed to reduce the computation time of the MILP approach. Section 4.6 illustrates the performance of the proposed approaches via a case study. A short summary is finally presented in Sect. 4.7.

4.2 Model Formulation

In this section, the formulation of operation of trains in the FBS system and in the MBS system are presented.

4.2.1 Train Dynamics

Although the model for the operation of trains has been described in the previous chapter, we here repeat the main equations that will be referred to in the remainder of this chapter. As shown in Chap. 3, the continuous-space model (3.14) and (3.15) of train operations can be discretized in space: the position horizon between two consecutive stations $[s_{\text{start}}, s_{\text{end}}]$ is split into N intervals and it is assumed that the track and train parameters as well as the traction or the breaking force can be considered as constant in each interval $[s_k, s_{k+1}]$ with length $\Delta s_k = s_{k+1} - s_k$, for $k = 1, 2, \ldots, N$. The discrete-space model is then transcribed into an piecewise affine model by approximating the nonlinear terms through piecewise affine functions. Furthermore, by applying the transformation properties described in Sect. 3.4.2, the piecewise affine model is formulated as the following mixed logical dynamic model:

$$\begin{aligned} x(k+1) = {} & A_k x(k) + B_k u(k) + C_{1,k}\delta(k) + C_{2,k}\delta(k+1) \\ & + D_{1,k} z(k) + D_{2,k} z(k+1) + e_k, \end{aligned} \tag{4.1}$$

$$R_{1,k}\delta(k) + R_{2,k}\delta(k+1) + R_{3,k}z(k) + R_{4,k}z(k+1) \le R_{5,k}u(k) + R_{6,k}x(k) + R_{7,k}, \tag{4.2}$$

where $x(k) = \left[E(k)\ t(k)\right]^T$, $\delta(\cdot)$ and $z(\cdot)$ are the binary variables and auxiliary variables introduced by the transformation, and (4.2) also includes the upper bounds and lower bounds constraints for $E(k)$, $t(k)$, and $u(k)$. The coefficient matrices in the mixed logical dynamic model are determined by the train model, the piecewise affine approximations, upper bounds and lower bounds constraints, etc.

4.2.2 *Operation of Trains in a Fixed Block Signaling System*

Figure 4.1 shows an example of a three-aspect FBS system with automatic train protection (ATP) speed codes. Later on, we will discuss the constraints caused by FBS system using this simple three-aspect signaling system. However, the methodology proposed in this chapter can be extended to other types of FBS systems.

In the FBS system given in Fig. 4.1, the speed code data consists of two parts [4], the authorized-speed code for this block and the target-speed code for the next block as illustrated in Fig. 4.1. When a train in Block 4 approaches to Signal 3, it will receive a $v_{\max}/v_{\text{yellow}}$ code, to indicate a permitted speed of $v_{\max}$ in this block and a target speed of v_{yellow} for the next. When the train enters Block 2, the received code is $v_{\text{yellow}}/v_{\min}$ because the next block (Block 1) is occupied by train 1, so the speed must be $v_{\min}$ (usually taking the value 0 m/s) by the time train reaches the end of Block 2. If the train attempts to pass the indication point for service braking distance before signal 1, the onboard equipment will cause an emergency brake application.

4.2.2.1 Minimum Headway for an FBS System

The minimum headway is introduced to ensure that the operation of a train is not impeded by signaling systems, i.e., a train's operation is not affected by the train in front. The minimum headway is defined as the minimum time separation between two successive trains at stations. For the undisturbed operation in FBS system, the minimum headway between the leading train and the following train can be computed as [5]

$$H_{\min,\text{FBS}} = \frac{L_{\text{a}}}{v_{\max}^{\text{F}}}\left[2+\text{INT}\left\{\frac{L_{\text{r}}^{\text{F}}+(v_{\max}^{\text{F}})^2/(2a_{\text{dec}}^{\text{F}})}{L_{\text{a}}}\right\}\right]+\frac{v_{\max}^{\text{F}}}{2a_{\text{dec}}^{\text{F}}}+t_{\text{d}}^{\text{L}}+\sqrt{\frac{2(L_{\text{t}}^{\text{L}}+L_{\text{s}})}{a_{\text{acc}}^{\text{L}}}}, \tag{4.3}$$

where INT$\{\cdot\}$ is a ceiling function that maps a real number to the least integer greater or equal to the argument, L_{a} is the block length, L_{r}^{F} is the distance that the following train will travel during the reaction time t_{r}^{F} of the driver and/or train control equipment of the following train, $v_{\max}^{\text{F}}$ is the maximum speed of the following train, $a_{\text{dec}}^{\text{F}}$ is the

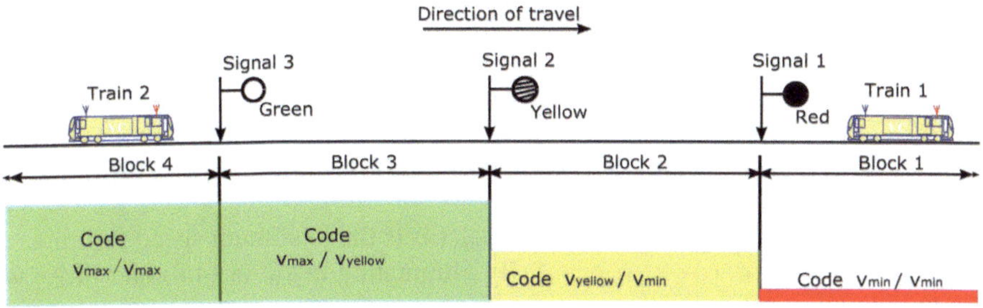

Fig. 4.1 Three-aspect fixed block signaling system with ATP speed codes

maximum service braking rate, $t_{\mathrm{d}}^{\mathrm{L}}$ is the station dwell time of the leading train, $L_{\mathrm{t}}^{\mathrm{L}}$ is the length of the leading train, L_{s} is the length of the secure section (a special section to protect the leading train), and $a_{\mathrm{acc}}^{\mathrm{L}}$ is the acceleration of the leading train.

4.2.2.2 The Constraints Caused by an FBS System

We assume that the total number of fixed block sections between two consecutive stations, i.e., in the interval $[s_{\mathrm{start}}, s_{\mathrm{end}}]$ is M. The index of block sections is denoted as m and the boundaries of the block sections are denoted as $s_{\mathrm{FB},m}$ with $m \in \{1, 2, \ldots, M\}$. Moreover, we assume that there exists an index $l_m \in \{1, 2, \ldots, N+1\}$ such that

$$s_{\mathrm{FB},m} = s_{l_m}, \tag{4.4}$$

and we define a piecewise constant function such that

$$\ell(k) = m, \quad \text{for} \quad l_m < k \leq l_{m+1}, \quad \text{for} \quad m \in \{1, 2, \ldots, M\}. \tag{4.5}$$

Then s_k is in the fixed block $(s_{l_{\ell(k)}}, s_{l_{\ell(k)+1}}]$. The constraints caused by the leading train in a three-aspect FBS system illustrated by Fig. 4.1 can be formulated as follows:

- If the following train and the leading train are in the same block section, i.e., $t^{\mathrm{F}}(s) \in \left(t^{\mathrm{L}}(s_{\mathrm{FB},m}), t^{\mathrm{L}}(s_{\mathrm{FB},m+1})\right]$ with $s \in [s_{\mathrm{FB},m}, s_{\mathrm{FB},m+1}]$, which is in fact not allowed by FBS systems, then the speed of the following train must equal to the minimum speed, i.e.,

$$v^{\mathrm{F}}(s) = v_{\min}. \tag{4.6}$$

- If the leading train is just one block section before the following train, i.e., $t^{\mathrm{F}}(s) \in \left(t^{\mathrm{L}}(s_{\mathrm{FB},m+1}), t^{\mathrm{L}}(s_{\mathrm{FB},m+2})\right]$ with $s \in [s_{\mathrm{FB},m}, s_{\mathrm{FB},m+1}]$, then the speed of the following train should be less than or equal to v_{Yellow} at position $s_{\mathrm{FB},m}$ and should be equal to $v_{\min}$ at position $s_{\mathrm{FB},m+1}$, i.e.,

$$\begin{aligned} v^{\mathrm{F}}(s_{\mathrm{FB},m}) &\leq v_{\mathrm{Yellow}}, \\ v^{\mathrm{F}}(s_{\mathrm{FB},m+1}) &= v_{\min}. \end{aligned} \tag{4.7}$$

The deceleration is assumed as a constant for the entire interval $[s_{\mathrm{FB},m}, s_{\mathrm{FB},m+1}]$. Based on the relationship among position, speed, and acceleration, we have

$$2a_{\mathrm{FB},m}(s_{\mathrm{FB},m+1} - s_{\mathrm{FB},m}) = v_{\min}^2 - v_{\mathrm{Yellow}}^2, \tag{4.8}$$

$$2a_{\mathrm{FB},m}(s - s_{\mathrm{FB},m}) = \bar{v}_{\mathrm{Yellow}}^2(s) - v_{\mathrm{Yellow}}^2, \tag{4.9}$$

where $a_{\mathrm{FB},m}$ is the deceleration and $\bar{v}_{\mathrm{Yellow}}(s)$ is the maximum speed for trains at position s for $s \in [s_{\mathrm{FB},m}, s_{\mathrm{FB},m+1}]$. By eliminating $a_{\mathrm{FB},m}$ in (4.8) and (4.9), we obtain

$$\bar{v}_{\text{Yellow}}(s) = \sqrt{v_{\text{Yellow}}^2 + (v_{\text{min}}^2 - v_{\text{Yellow}}^2)\frac{s - s_{\text{FB},m}}{s_{\text{FB},m+1} - s_{\text{FB},m}}}, \tag{4.10}$$

where $\bar{v}_{\text{Yellow}}(\cdot)$ is a function only depending on s. Therefore, in this case we have the constraint

$$v^{\text{F}}(s) \leq \bar{v}_{\text{Yellow}}(s). \tag{4.11}$$

- If the leading train is two blocks before the follower, i.e., $t^{\text{F}}(s) \in \left(t^{\text{L}}(s_{\text{FB},m+2}), t^{\text{L}}(s_{\text{FB},m+3})\right]$ with $s \in [s_{\text{FB},m}, s_{\text{FB},m+1}]$, then the speed of the following train at positions $s_{\text{FB},m}$ and $s_{\text{FB},m+1}$ should be less than or equal to v_{max} and v_{Yellow}, respectively, i.e.,

$$\begin{aligned} v^{\text{F}}(s_{\text{FB},m}) &\leq v_{\text{max}}, \\ v^{\text{F}}(s_{\text{FB},m+1}) &\leq v_{\text{Yellow}}. \end{aligned} \tag{4.12}$$

Similarly as $\bar{v}_{\text{Yellow}}(\cdot)$, we can obtain

$$\bar{v}_{\text{max}}(s) = \sqrt{v_{\text{max}}^2 + (v_{\text{Yellow}}^2 - v_{\text{max}}^2)\frac{s - s_{\text{FB},m}}{s_{\text{FB},m+1} - s_{\text{FB},m}}}, \tag{4.13}$$

where $\bar{v}_{\text{max}}(s)$ is the maximum speed for trains at position s. Note that $\bar{v}_{\text{max}}(s)$ only depends on s. Therefore, we in this case have the constraint

$$v^{\text{F}}(s) \leq \bar{v}_{\text{max}}(s). \tag{4.14}$$

4.2.3 *Operation of Trains in a Moving Block Signaling System*

In a pure MBS system,[1] the distance between two consecutive trains should be larger than the minimum distance at any time as shown in Fig. 4.2. The minimum distance between two successive trains is basically the instantaneous braking distance required by the following train plus a safety margin. As mentioned in Sect. 2.1.2, in practice the minimum distance in the MBS system is larger than that defined in theory, because the driver or the automatic train control system needs time to react to situations. The distance between the leading train and the following train in an MBS system should satisfy [4]

$$s^{\text{L}}(t) - s^{\text{F}}(t) \geq L_{\text{r}}^{\text{F}} + (v^{\text{F}}(t))^2/(2a_{\text{dec}}^{\text{F}}) + S_{\text{SM}} + L_{\text{t}}^{\text{L}}, \tag{4.15}$$

[1] As stated in Chap. 2, there exist four MBS schemes: moving space blocking signaling, moving time block signaling, pure MBS, and relative MBS. In this book, we only consider the pure MBS system, so the MBS system later on refers to the pure MBS system.

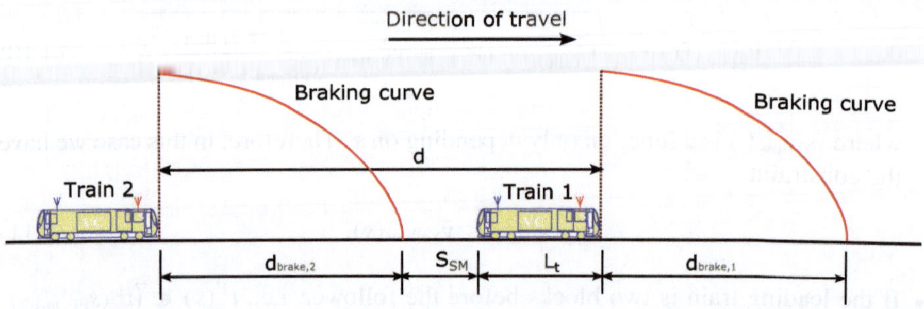

Fig. 4.2 A moving block signaling system

where $s^{\mathrm{L}}(t)$ and $s^{\mathrm{F}}(t)$ are the positions of the front of the leading train and the following train at time t, $v^{\mathrm{F}}(t)$ is the speed of the following train, $a^{\mathrm{F}}_{\mathrm{dec}}$ is the maximal deceleration of the following train, S_{SM} is the safety margin distance, $L^{\mathrm{F}}_{\mathrm{r}}$ is the distance that the following train will travel during the reaction time $t^{\mathrm{F}}_{\mathrm{r}}$ of the driver and/or train equipment of the following train, and $L^{\mathrm{L}}_{\mathrm{t}}$ is the length of the leading train. The value of the reaction time could be obtained from historical data.

4.2.3.1 Minimum Headway for a MBS System

The minimum distance between two successive trains (4.15) can be recast as the minimum time difference of two successive trains

$$t^{\mathrm{F}}(s) - t^{\mathrm{L}}(s) \geq t^{\mathrm{F}}_{\mathrm{r}} + t^{\mathrm{F}}_{\mathrm{dec}}(s) + t^{\mathrm{F}}_{\mathrm{safe}}(s), \tag{4.16}$$

where $t^{\mathrm{L}}(s)$ and $t^{\mathrm{F}}(s)$ are the time instants at which the front of the leading train and the following train pass position s, respectively. The deceleration time of the following train $t^{\mathrm{F}}_{\mathrm{dec}}(s)$ and the time margin $t^{\mathrm{F}}_{\mathrm{safe}}(s)$ caused by the safe margin distance and the train length can be computed as

$$t^{\mathrm{F}}_{\mathrm{dec}}(s) = v^{\mathrm{F}}(s)/a^{\mathrm{F}}_{\mathrm{dec}}, \tag{4.17}$$

$$t^{\mathrm{F}}_{\mathrm{safe}}(s) = (S_{\mathrm{SM}} + L^{\mathrm{L}}_{\mathrm{t}})/v^{\mathrm{F}}(s), \tag{4.18}$$

where $v^{\mathrm{F}}(s)$ is the speed of the following train at position s.

In order to ensure that near stations the operation of a train is not impeded by MBS systems, i.e., a train's operation is not then affected by the (in principle slowly moving or stopped) train in front, the minimum headway is introduced. The minimum headway is the minimum time separation between successive trains at train stations. The minimum headway of MBS systems is calculated as [4]

$$H_{\mathrm{min,MBS}} = t^{\mathrm{L}}_{\mathrm{d}} + t_{\mathrm{in-out}} = t^{\mathrm{L}}_{\mathrm{d}} + t^{\mathrm{F}}_{\mathrm{r}} + t^{\mathrm{F}}_{\mathrm{dec,max}} + t^{\mathrm{L}}_{\mathrm{safe}}, \tag{4.19}$$

with the run-in/run-out time $t_{\text{in-out}} = t_{\text{r}}^{\text{F}} + t_{\text{dec,max}}^{\text{F}} + t_{\text{safe}}^{\text{L}}$, where $t_{\text{dec,max}}^{\text{F}}$ is the time it takes the following train to come to a full stop when it is running at its maximum speed, i.e., $t_{\text{dec,max}}^{\text{F}} = v_{\text{max}}^{\text{F}}/a_{\text{b}}^{\text{F}}$, and the run-out time $t_{\text{safe}}^{\text{L}}$ is the time that the leading train needs to completely clear the secure section (i.e., a special section to protect the leading train), if present, and including a safety margin, i.e., $t_{\text{safe}}^{\text{L}} = \sqrt{2(S_{\text{SM}} + L_{\text{t}}^{\text{L}} + L_{\text{s}})/a_{\text{acc}}^{\text{L}}}$. The acceleration of the leading train $a_{\text{acc}}^{\text{L}}$ is usually considered as a constant value for the minimum headway calculation [5].

4.2.3.2 The Constraints Caused by an MBS System

The constraints caused by the leading train in MBS system are different in open track area and in station area. In open track area, the minimum time difference between the leading train and the following train should satisfy (4.16). In station area, the minimum time distance should be larger than the minimum headway defined in (4.19), i.e.,

$$t^{\text{F}}(s) - t^{\text{L}}(s) \geq H_{\text{min,MBS}}. \tag{4.20}$$

4.3 Mathematical Formulation of the Multiple Trains Trajectory Planning Problem

For simplicity, we consider the optimal trajectory planning problem for two trains. However, the solution approaches can be extended to multiple trains. The trajectory planning problem for two trains (i.e., the leading train and the following train) can be formulated as follows:

$$J = \int_{s_{\text{start}}}^{s_{\text{end}}} \max\left(0, u^{\text{L}}(s)\right)\mathrm{d}s + \int_{s_{\text{start}}}^{s_{\text{end}}} \max\left(0, u^{\text{F}}(s)\right)\mathrm{d}s, \tag{4.21}$$

subject to

$$\begin{aligned}
&u_{\min} \leq u^{\text{L}}(s) \leq u_{\max}, \quad u_{\min} \leq u^{\text{F}}(s) \leq u_{\max}, \quad \forall s \in [s_{\text{start}}, s_{\text{end}}]\\
&0 < \tilde{E}^{\text{L}}(s) \leq \tilde{E}_{\max}^{\text{L}}(s), \quad 0 < \tilde{E}^{\text{F}}(s) \leq \tilde{E}_{\max}^{\text{F}}(s), \quad \forall s \in [s_{\text{start}}, s_{\text{end}}]\\
&\tilde{E}^{\text{L}}(s_{\text{start}}) = \tilde{E}_{\text{start}}^{\text{L}}, \quad \tilde{E}^{\text{F}}(s_{\text{start}}) = \tilde{E}_{\text{start}}^{\text{F}},\\
&\tilde{E}^{\text{L}}(s_{\text{end}}) = \tilde{E}_{\text{end}}^{\text{L}}, \quad \tilde{E}^{\text{F}}(s_{\text{end}}) = \tilde{E}_{\text{end}}^{\text{F}},\\
&t^{\text{L}}(s_{\text{start}}) = T_{\text{start}}^{\text{L}}, \quad t^{\text{F}}(s_{\text{start}}) = T_{\text{start}}^{\text{F}},\\
&t^{\text{L}}(s_{\text{end}}) = T_{\text{end}}^{\text{L}}, \quad t^{\text{F}}(s_{\text{end}}) = T_{\text{end}}^{\text{F}},
\end{aligned} \tag{4.22}$$

the train model constraints (4.1) and (4.2) for the leading train and the following train, and the constraints caused by the signaling systems in Sects. 4.2.2 and 4.2.3 (i.e.,

(4.6), (4.7), (4.11), (4.12), (4.14), (4.16), and (4.20)), where the objective function J is the energy consumption for these two trains without regenerative braking; $E^{\mathrm{L}}_{\max}(\cdot)$ is equal to $0.5 \cdot (V^{\mathrm{L}}_{\max}(\cdot))^2$ with $V^{\mathrm{L}}_{\max}(\cdot)$ the maximum allowable velocity, which depends on the train characteristics and line conditions, and as such it is usually a piecewise constant function of the coordinate s [6, 7]; s_{start}, $\tilde{E}^{\mathrm{L}}(s_{\mathrm{start}})$, and $t(s_{\mathrm{start}})$ are the position, the kinetic energy per mass, and the departure time for the leading train at the beginning of the route; s_{end}, $\tilde{E}^{\mathrm{L}}(s_{\mathrm{end}})$, and $t^{\mathrm{L}}(s_{\mathrm{end}})$ are the position, the kinetic energy per mass, and the arrival time for the leading train at the end of the route. The notation definitions for the following train are similar. It is assumed that the unit kinetic energy $\tilde{E}^{\mathrm{L}}(s) > 0$ and $\tilde{E}^{\mathrm{F}}(s) > 0$ for the leading train and the following train, which means the train's speed is always strictly larger than zero, i.e., the train travels nonstop [6]. This assumption is nonrestrictive in practice because the initial start and terminal stop can be modeled by small nonzero velocities. Furthermore, in principle the traffic management system does not plan stops intentionally at an intermediate point of the trip.

In addition, the constraints caused by the FBS system or the MBS system proposed in Sects. 4.2.2 and 4.2.3 should also be included in the trajectory planning problem.

Remark 4.1 The ideas on solving the train scheduling problem in a rolling horizon way in Sect. 7.5 could be adopted to the train trajectory planning for multiple trains in this chapter.

4.4 Solution Approaches

Now two solution approaches, i.e., the greedy approach and the simultaneous approach, are proposed for solving the optimal control problem for multiple trains under an FBS and or an MBS system.

4.4.1 *Greedy Approach*

In the greedy approach, the leading train's trajectory is first determined and then the trajectory of the following train is optimized based on the results of the leading train. For the trajectory planning problem of the leading train, the objective is only part of (4.21), i.e.,

$$J^{\mathrm{L}} = \int_{s_{\mathrm{start}}}^{s_{\mathrm{end}}} \max\left(0, u^{\mathrm{L}}(s)\right) \mathrm{d}s, \tag{4.23}$$

and the constraints are those related to the leading train. The constraints pertaining to both the leading train and the following train are discarded for the trajectory planning for the leading train. The optimization problem of the leading train is the same as the problem solved in Chap. 3, and it can be solved using the pseudospectral method or the MILP approach.

The optimal trajectory planning problem for the following train is similar with that of the leading train. When optimizing the trajectory of the following train, the trajectory of the leader train is already fixed and the value of J^{L} is also fixed. The objective for the planning problem of the following train is

$$J^{\mathrm{F}} = \int_{s_{\mathrm{start}}}^{s_{\mathrm{end}}} \max\big(0, u^{\mathrm{F}}(s)\big)\mathrm{d}s. \tag{4.24}$$

The constraints do not only include those in (4.22) related to the following train, but also the constraints caused by the FBS or the MBS system as presented in Sects. 4.2.2 or 4.2.3.

In the pseudospectral method, the constraints caused by the signaling system can be easily formulated as the path constraints (see Sect. 3.4.1 for more information). However, for the MILP approach, we need to approximate the nonlinear constraints into linear constraints using the transformation properties in Sect. 3.4.2. The details for the transformation process of the FBS constraints and MBS constraints will be given in Sect. 4.5.

4.4.2 Simultaneous Approach

The simultaneous approach optimizes the trajectories of the leading train and the following train simultaneously. When optimizing the trajectories of multiple trains at the same time, the model for each train is a model of the form (3.1). The trajectories of the leading train and the following train are obtained at the same time and the value of the objective function (4.21) is minimized by solving an optimization problem involving multiple trains. This problem is similar with the trajectory planning problem for a single train and therefore it can also be solved by the pseudospectral method or the MILP approach. The constraints caused by the FBS system and MBS system can be handled similarly as in the greedy approach.

However, compared to the case of a single train, the number of the state variables and constraints of the problem for multiple trains increase linearly with the number of trains. Therefore, the size of the optimal trajectory planning problem for multiple trains is much bigger than the problem for a single train and the computation time of the bigger problem will be much longer. However, since we are now optimizing the train trajectories of two trains at the same time instead of optimizing them one by one, the control performance will in general be better than that of the greedy approach.

Remark 4.2 For the trajectory planning for large number of trains, the simultaneous approach will become slow. So distributed optimization approaches [8] can be applied, where the trajectory of each train could be calculated separately with consideration of constraints caused by other trains, then these trains negotiate with each other and finally converge to a global equilibrium.

4.5 Mixed Logical Dynamic Formulation for Signaling System Constraints

As shown in previous chapter, the optimal control problem can be recast as an MILP problem of the following form:

$$\min_{\tilde{V}} \quad C_J^T \tilde{V}, \tag{4.25}$$

subject to

$$\begin{aligned} F_1\tilde{V} &\leq F_2 x(1) + f_3, \\ F_4\tilde{V} &= F_5 x(1) + f_6. \end{aligned} \tag{4.26}$$

In order to include the constraints caused by the FBS system and MBS system (shown in Sects. 4.2.2 and 4.2.3) in the MILP problem, these constraints should be transformed into linear constraints by using the transformation properties in Sect. 3.4.2. The transformation of these constraints is given as below next.

4.5.1 Multiple Trains Under Fixed Block Signaling System

The constraints caused by the leading train in the FBS system are first discretized at each grid point s_k for $k = 1, 2, \ldots, N+1$. These logical constraints are then transformed into linear constraints, which can be easily included in the MILP approach.

4.5.1.1 Discretizing the FBS System Constraints

As described in Sect. 4.2.2, the fixed block sections are indexed by m with $m \in \{1, 2, \ldots, M\}$ and the boundaries for the block sections are denoted $s_{\mathrm{FB},m}$ and grid point s_k is in the fixed block $(s_{l_{\ell(k)}}, s_{l_{\ell(k)+1}}]$. The constraints caused by the leading train in a three-aspect FBS system shown in Fig. 4.1 can be transformed at each discrete point s_k as follows:

- If the following train and the leading train are in the same block section, i.e., $t^{\mathrm{F}}(k) \in \left(t^{\mathrm{L}}(l_{\ell(k)}), t^{\mathrm{L}}(l_{\ell(k)+1})\right]$, then the speed of the following train must equal to the minimum speed, i.e.,

$$v^{\mathrm{F}}(k) = v_{\min}. \tag{4.27}$$

- If the leading train is one block section before the following train, i.e., $t^{\mathrm{F}}(k) \in \left(t^{\mathrm{L}}(l_{\ell(k)+1}), t^{\mathrm{L}}(l_{\ell(k)+2})\right]$, then the speed of the following train at positions $s_{l_{\ell(k)}}$ and $s_{l_{\ell(k)+1}}$ should be less than or equal to V_{Yellow} and equal to $V_{\min}$, respectively, i.e.,

$$\begin{aligned} v^{\mathrm{F}}_{l_{\ell(k)}} &\leq v_{\mathrm{Yellow}}, \\ v^{\mathrm{F}}_{l_{\ell(k)+1}} &= v_{\min}. \end{aligned} \tag{4.28}$$

For the grid points $s_k \in (s_{l_{\ell(k)}}, s_{l_{\ell(k)+1}})$, we have

$$v^{\mathrm{F}}(k) \leq \bar{v}_{\mathrm{Yellow},k}, \tag{4.29}$$

where

$$\bar{v}_{\mathrm{Yellow},k} = \sqrt{v^2_{\mathrm{Yellow}} + (v^2_{\min} - v^2_{\mathrm{Yellow}})\frac{s_k - s_{l_{\ell(k)}}}{s_{l_{\ell(k)+1}} - s_{l_{\ell(k)}}}}. \tag{4.30}$$

- If the leading train is two blocks before the follower, i.e., $t^{\mathrm{F}}(k) \in \left(t^{\mathrm{L}}(l_{\ell(k)+2}), t^{\mathrm{L}}(l_{\ell(k)+3})\right]$, then the speed of the following train at positions $s_{l_{\ell(k)}}$ and $s_{l_{\ell(k)+1}}$ should be less than or equal to $V_{\max}$ and V_{Yellow}, respectively, i.e.,

$$\begin{aligned} v^{\mathrm{F}}_{l_{\ell(k)}} &\leq v_{\max}, \\ v^{\mathrm{F}}_{l_{\ell(k)+1}} &\leq v_{\mathrm{Yellow}}. \end{aligned} \tag{4.31}$$

For the grid points $s_k \in (s_{l_{\ell(k)}}, s_{l_{\ell(k)+1}})$, we have

$$v^{\mathrm{F}}(k) \leq \bar{v}_{\max,k}, \tag{4.32}$$

where

$$\bar{v}_{\max,k} = \sqrt{v^2_{\max} + (v^2_{\mathrm{Yellow}} - v^2_{\max})\frac{s_k - s_{l_{\ell(k)}}}{s_{l_{\ell(k)+1}} - s_{l_{\ell(k)}}}}. \tag{4.33}$$

4.5.1.2 Considering the FBS Constraints into the MILP Approach

In order to transform the above logical constraints into linear constraints, the following binary variables are introduced:

$$\begin{aligned} [t^{\mathrm{F}}(k) \leq t^{\mathrm{L}}(l_{\ell(k)+1})] &\Leftrightarrow [\delta_1(k) = 1], \\ [t^{\mathrm{F}}(k) \leq t^{\mathrm{L}}(l_{\ell(k)+2})] &\Leftrightarrow [\delta_2(k) = 1], \\ [t^{\mathrm{F}}(k) \leq t^{\mathrm{L}}(l_{\ell(k)+3})] &\Leftrightarrow [\delta_3(k) = 1]. \end{aligned} \tag{4.34}$$

Note that an extra constraint is needed, i.e.,

$$t^{\mathrm{F}}(k) > t^{\mathrm{L}}(k), \tag{4.35}$$

because the passing time of the following train at position s_k should be larger than that of the leading train in order to avoid the collision of trains. In addition, based on the definition of $\delta_1(k)$, $\delta_2(k)$, and $\delta_3(k)$, the following logical conditions are satisfied

$$\begin{aligned}&[\delta_1(k)=1] \Leftrightarrow [\delta_2(k)=\delta_3(k)=1],\\&[\delta_2(k)=1] \Leftrightarrow [\delta_3(k)=1].\end{aligned} \tag{4.36}$$

The constraints caused by the leading train in the three-aspect FBS system can then be reformulated as:

$$\delta_1(k)v^{\mathrm{F}}(k) \leq v_{\min}, \tag{4.37}$$

$$(1-\delta_1(k))\delta_2(k)v^{\mathrm{F}}(k) \leq \bar{v}_{\mathrm{Yellow},k}, \tag{4.38}$$

$$(1-\delta_2(k))\delta_3(k)v^{\mathrm{F}}(k) \leq \bar{v}_{\max,k}, \tag{4.39}$$

where '$\leq$' is used in (4.37) instead of '$=$' in (4.27) because $\delta_1(k)v^{\mathrm{F}}(k)$ is equal to 0 and is not equal to $v_{\min}$ when $\delta_1(k)=0$.

By defining $\tilde{M}_i = T^{\mathrm{F}}_{\max} - t^{\mathrm{L}}(l_{\ell(k)+i})$ that is larger than $\max(t^{\mathrm{F}}(k) - t^{\mathrm{L}}(l_{\ell(k)+i}))$, $\tilde{m}_i = \min(t^{\mathrm{F}}(k) - t^{\mathrm{L}}(l_{\ell(k)+i}))$ that is larger than $T^{\mathrm{F}}_{\min} - t^{\mathrm{L}}(l_{\ell(k)+i})$, and by applying transformation property (3.35) the logical constraints (4.34) can be shown to be equivalent to the following inequalities:

$$\begin{aligned}&t^{\mathrm{F}}(k) - t^{\mathrm{L}}(l_{\ell(k)+i}) \leq \tilde{M}_i(1-\delta_i(k)),\\&t^{\mathrm{F}}(k) - t^{\mathrm{L}}(l_{\ell(k)+i}) \geq \varepsilon + (\tilde{m}_i - \varepsilon)\delta_i(k),\end{aligned} \tag{4.40}$$

for $i = 1, 2, 3$, where $T^{\mathrm{F}}_{\max}$ is the arrival time of the following train at the final destination, $T^{\mathrm{F}}_{\min}$ the departure time of the following train, and ε is the machine precision. In addition, we define binary variables $\delta_4(k) = \delta_1(k)\delta_2(k)$ and $\delta_5(k) = \delta_2(k)\delta_3(k)$ to deal with the nonlinear terms $\delta_1(k)\delta_2(k)$ and $\delta_2(k)\delta_3(k)$ in (4.38)–(4.39) respectively. According to the transformation properties in Sect. 3.4.2, the definitions of $\delta_4(k)$ and $\delta_5(k)$ are equivalent to linear constraints of the form (3.36). In addition, auxiliary variables $z^{\mathrm{F}}_i(k)$ are introduced to deal with the nonlinear terms $\delta_i(k)v^{\mathrm{F}}(k)$, which is defined as

$$z^{\mathrm{F}}_i(k) = \delta_i(k)v^{\mathrm{F}}(k), \quad \text{for} \quad i = 1, 2, 3, 4, 5. \tag{4.41}$$

This definition is equivalent to linear constraints of the form (3.37). The constraints caused by the leading train in fixed block systems can thus be formulated into the MILP problem setting.

In the greedy approach, first the trajectory of the leading train is determined by solving the MILP problem. Next, the optimal control problem of the following train is solved, which is similar to the one of Chap. 3 but with the extra constraints caused by the leading train in FBS system. The coefficient matrices in the mixed logical dynamic model (4.1)–(4.2) are determined by the following train. Since the trajectory of the leading train is known, $t^{\mathrm{L}}(l_{\ell(k)+i})$ is also known. Therefore, $\tilde{M}_i$ and $\tilde{m}_i$ are constants and (4.40) is a system of linear constraints.

When optimizing the trajectories of multiple trains simultaneously, the models for these two trains are determined by a model of the form (4.1) to (4.2). The optimal control problem of these two successive trains can also be rewritten in the form of the MILP problem (4.25)–(4.26) but including the model and constraints of each train and the constraints caused by the FBS system. However, $t^{\mathrm{L}}(l_{\ell(k)+i})$ is now also a variable in this case since the leading train's trajectory also has to be optimized. The constraints (4.40) can be rewritten as

$$\begin{aligned}
&(T^{\mathrm{F}}_{\max} - t^{\mathrm{L}}(l_{\ell(k)+i}))\delta_i(k) \leq -t^{\mathrm{F}}(k) + T^{\mathrm{F}}_{\max}, \\
&(T^{\mathrm{F}}_{\min} - t^{\mathrm{L}}(l_{\ell(k)+i}) - \varepsilon)\delta_i(k) \leq t^{\mathrm{F}}(k) - \varepsilon - t^{\mathrm{L}}(l_{\ell(k)+i}).
\end{aligned} \tag{4.42}$$

In order to deal with the nonlinear terms in (4.42), we define

$$z^{\mathrm{L}}_i(k) = t^{\mathrm{L}}(l_{\ell(k)+i})\delta_i(k), \quad \text{for} \quad i = 1, 2, 3. \tag{4.43}$$

Similar to $z^{\mathrm{F}}_i(k)$, (4.43) is equivalent to linear constraints according to the transformation properties in Sect. 3.4.2.

4.5.2 *Multiple Trains Under Moving Block Signaling System*

Just as the constraints caused by the FBS system, the constraints caused by the MBS system are first discretized at the grid points s_k, $k = 1, \ldots, N + 1$. Next, these constraints are approximated by linear constraints, which can easily be included in the MILP formulation (4.25)–(4.26).

4.5.2.1 Discretizing the MBS Constraints

We discretize the constraint (4.16) caused by the MBS system at the grid points s_k as

$$t^{\mathrm{F}}(k) \geq t^{\mathrm{L}}(k) + t^{\mathrm{F}}_{\mathrm{r}} + t^{\mathrm{F}}_{\mathrm{dec}}(k) + t^{\mathrm{F}}_{\mathrm{safe}}(k), \quad \text{for} \quad k = 1, 2, \ldots, N, \tag{4.44}$$

$$t^{\mathrm{F}}(k) \geq t^{\mathrm{L}}(k) + t^{\mathrm{F}}_{\mathrm{r}} + t^{\mathrm{F}}_{\mathrm{dec,max}} + t^{\mathrm{L}}_{\mathrm{d}} + t^{\mathrm{L}}_{\mathrm{safe}}, \quad \text{for} \quad k = N + 1. \tag{4.45}$$

In addition, some intermediate constraints are introduced to ensure that the points between the grid points also satisfy the constraints caused by the MBS system. According to (4.16), we obtain the following constraint for each $s \in [s_k, s_{k+1}]$ as follows:

$$t^{\mathrm{F}}(s) - t^{\mathrm{F}}_{\mathrm{r}} - t^{\mathrm{F}}_{\mathrm{dec}}(s) - t^{\mathrm{F}}_{\mathrm{safe}}(s) \geq t^{\mathrm{L}}(s). \tag{4.46}$$

If we assume the left-hand side of (4.46) to be an affine function in the interval $[s_k, s_{k+1}]$, then we can add the following constraints:

$$(1-\alpha)(t^{\mathrm{F}}(k) - t_{\mathrm{r}}^{\mathrm{F}} - t_{\mathrm{dec}}^{\mathrm{F}}(k) - t_{\mathrm{safe}}^{\mathrm{F}}(k)) + \alpha(t^{\mathrm{F}}(k+1) - t_{\mathrm{r}}^{\mathrm{F}} - t_{\mathrm{dec}}^{\mathrm{F}}(k+1) - t_{\mathrm{safe}}^{\mathrm{F}}(k+1)) \\ \geq t^{\mathrm{L}}(s + \alpha \Delta s_k), \tag{4.47}$$

for some values α in a finite subset $S_\alpha \in [0, 1)$, e.g., $S_\alpha = \{0.1, 0.2, \ldots, 0.9\}$, where $t^{\mathrm{L}}(s + \alpha\Delta s_k)$ is known if the optimal trajectory of the leading train is fixed. Note that for $\alpha = 0$ and $\alpha = 1$, the constraint (4.44) is retrieved (except if $k = N - 1$). However, when we solve the trajectory planning problem for multiple trains using the simultaneous approach, the leading train's trajectory is not known beforehand. So the term $t^{\mathrm{L}}(s + \alpha\Delta s_k)$ is unknown. If we assume the right-hand side of (4.46) is also an affine function, i.e., $t^{\mathrm{L}}(s + \alpha\Delta s_k) = (1-\alpha)t^{\mathrm{L}}(k) + \alpha t^{\mathrm{L}}(k+1)$, then it is sufficient to check (4.46) at the points k and $k+1$ (i.e., for $\alpha = 0$ and $\alpha = 1$), since due to linearity (4.46) will then also automatically be satisfied for all intermediary points.

4.5.2.2 Considering the MBS Constraints into the MILP Approach

Note that the constraints (4.44), (4.45), and (4.47) are linear in $t^{\mathrm{L}}(k)$, $t^{\mathrm{L}}(k+1)$, and $t^{\mathrm{F}}(k)$. However, they are nonlinear in $v^{\mathrm{F}}(k)$ and $v^{\mathrm{F}}(k+1)$ since the time safety margin (4.18) is a nonlinear function of the following train's velocity $v^{\mathrm{F}}(k)$. Furthermore, the kinetic energy per mass $E^{\mathrm{F}}(k)$ is one of the states instead of $v^{\mathrm{F}}(k)$ with $E^{\mathrm{F}}(k) = 0.5(v^{\mathrm{F}}(k))^2$ (cf. Sect. 4.2.1). Therefore, both the braking time $t_{\mathrm{dec}}^{\mathrm{F}}(k)$ and the safe time margin $t_{\mathrm{safe}}^{\mathrm{F}}(k)$ are nonlinear functions of $E^{\mathrm{F}}(k)$, where

$$t_{\mathrm{dec}}^{\mathrm{F}}(k) = \frac{1}{a_{\mathrm{dec}}^{\mathrm{F}}}\sqrt{2E^{\mathrm{F}}(k)}, \tag{4.48}$$

and

$$t_{\mathrm{safe}}^{\mathrm{F}}(k) = (S_{\mathrm{SM}} + L_{\mathrm{t}}^{\mathrm{L}})\frac{1}{\sqrt{2E^{\mathrm{F}}(k)}}. \tag{4.49}$$

The nonlinear functions $f_1(\cdot) : E^{\mathrm{F}} \to \sqrt{2E^{\mathrm{F}}}$ and $f_2(\cdot) : E^{\mathrm{F}} \to \frac{1}{\sqrt{2E^{\mathrm{F}}}}$ could be approximated by piecewise affine functions as follows (see Chap. 3 for more details about piecewise affine approximation):

$$f_{1,\mathrm{PWA}}(E^{\mathrm{F}}(k)) = \begin{cases} \alpha_1 E^{\mathrm{F}}(k) + \beta_1 \text{ for } E_{\min} \leq E^{\mathrm{F}}(k) < E_1, \\ \alpha_2 E^{\mathrm{F}}(k) + \beta_2 \text{ for } E_1 \leq E^{\mathrm{F}}(k) \leq E_{\max}, \end{cases} \tag{4.50}$$

$$f_{2,\mathrm{PWA}}(E^{\mathrm{F}}(k)) = \begin{cases} \lambda_1 E^{\mathrm{F}}(k) + \mu_1 \text{ for } E_{\min} \leq E^{\mathrm{F}}(k) < E_1, \\ \lambda_2 E^{\mathrm{F}}(k) + \mu_2 \text{ for } E_1 \leq E^{\mathrm{F}}(k) \leq E_{\max}, \end{cases} \tag{4.51}$$

with optimized parameters α_1, α_2, β_1, β_2, λ_1, λ_2, μ_1, μ_2, and E_1. For more details of this transformation into piecewise affine functions, see [9]. Now the constraint (4.44) can be approximated as the following linear constraint:

$$\begin{aligned} t^{\mathrm{F}}(k) \geq\, & t^{\mathrm{L}}(k) + t_{\mathrm{r}}^{\mathrm{F}} + \frac{1}{a_{\mathrm{dec}}^{\mathrm{F}}}(\alpha_1 E^{\mathrm{F}}(k) + \beta_1) \\ & + (S_{\mathrm{SM}} + L_{\mathrm{t}}^{\mathrm{L}})(\lambda_1 E^{\mathrm{F}}(k) + \mu_1), \text{ if } E_{\min} \leq E^{\mathrm{F}}(k) < E_1, \end{aligned} \tag{4.52}$$

$$\begin{aligned} t^{\mathrm{F}}(k) \geq\, & t^{\mathrm{L}}(k) + t_{\mathrm{r}}^{\mathrm{F}} + \frac{1}{a_{\mathrm{dec}}^{\mathrm{F}}}(\alpha_2 E^{\mathrm{F}}(k) + \beta_2) \\ & + (S_{\mathrm{SM}} + L_{\mathrm{t}}^{\mathrm{L}})(\lambda_2 E^{\mathrm{F}}(k) + \mu_2), \text{ if } E_1 \leq E^{\mathrm{F}}(k) \leq E_{\max}. \end{aligned} \tag{4.53}$$

Similarly, the constraints (4.45) and (4.46) can also be written as linear constraints. These approximated linear constraints caused by the MBS system can be easily included in the MILP approach and we still get an MILP problem.

The greedy approach and the simultaneous approach presented in Sect. 4.4 can be applied for trajectory planning of multiple trains in MBS systems. In the greedy approach $t^{\mathrm{L}}(k)$ and $t^{\mathrm{L}}(s + \alpha\Delta s_k)$ are known for the trajectory planning problem for the following train since the trajectory of the leading train is known by the zone controller or the following train. The trajectory planning problem for the following train is similar to the one of Chap. 3 but with the MBS constraints of Sect. 4.5.2. When optimizing the trajectories of multiple trains at the same time, the model for each of these trains is determined by a model of the form (4.1) to (4.2). The optimal control problem of these successive trains can also be rewritten in the form of the MILP problem (4.25)–(4.26).

4.5.3 Extension: Mode Vector Constraints

The computation time needed for solving an MILP problem for a single train is usually small if we take a small value for N. However, the computation time increases quickly with the value of N and the number of trains considered in the MILP problem for multiple trains. In the worst case, the computation time grows exponentially with the number of integer variables. In order to solve the MILP problem for multiple trains in a reasonable time, we introduce the so-called mode vector constraints, which have already been applied to sewer networks [10].

A mode of the mixed logical dynamic model for a single train refers to a specific value of the binary vector $\delta^i = \left[\delta_1^i \; \delta_2^i \; \ldots \; \delta_{M_i}^i\right]^T$, where i represents the ith train and M_i denotes the dimension of the binary vector of the ith train (see the MILP formulation (4.25)–(4.26)). Furthermore, a mode vector is defined as a tuple of binary vectors for each train considered in the MILP problem, i.e., $\Delta = (\delta^{1T}, \delta^{2T}, \ldots, \delta^{IT})^T$, where I is the number of trains considered in the problem. Let $\bar{\Delta} = (\bar{\delta}^{1T}, \bar{\delta}^{2T}, \ldots, \bar{\delta}^{IT})^T$ be a reference mode vector of the MILP problem for multiple trains. Note however that this comes at a cost of reduced optimality; the optimality can next be improved again by solving the full I trains MILP problem with the mode constraints. The mode vector constraints can be defined as

$$\sum_{m=1}^{M_i} |\bar{\delta}^i_m - \delta^i_m| \leq D_i \quad \text{for} \quad i = 1, 2, \ldots, I, \tag{4.54}$$

or as

$$\sum_{i=1}^{I} \sum_{m=1}^{M_i} |\bar{\delta}^i_m - \delta^i_m| \leq D, \tag{4.55}$$

where D_i and D are preselected bounds (a nonnegative integer value) on the number of 0–1switches (or vice versa) in the entires of the updated mode vector with respect to the reference mode vector. Note that the mode vector constraints (4.54) and (4.55) can be recast as linear constraints by introducing some auxiliary variables to deal with the absolute values $|\bar{\delta}^i_m - \delta^i_m|$ for $i = 1, 2, \ldots, I$ and $m = 1, 2, \ldots, M_i$. Furthermore, the mode vector constraints can be seen as the Hamming distance between Δ and $\bar{\Delta}$ if we would expand Δ and $\bar{\Delta}$ into binary strings [10].

An important practical problem is to find a reference mode vector $\bar{\Delta}$. It is stated in [11] that physics or heuristic knowledge of the system can often be used to find $\bar{\Delta}$ that fulfills the physical constraints of the system. A good candidate for the reference mode vector can be obtained by solving the optimal control problem for multiple trains sequentially since the computation time of the solving I single-train MILP problems will be much less than solving the full I-trains MILP problem all at once.

By using the MILP approach with mode vector constraints, the computation time can be reduced significantly. A case study illustrate this is presented in [3], see there for more information.

4.6 Case Study

In order to illustrate the performance of the proposed greedy and simultaneous approaches for the optimal trajectory planning for multiple trains under an FBS system and an MBS system, a part of the Beijing Yizhuang subway line is used as a test case study.

4.6.1 Set-Up

The performance of the MILP approach is compared with the widely used pseudo-spectral method and for both approaches we consider both the greedy and the

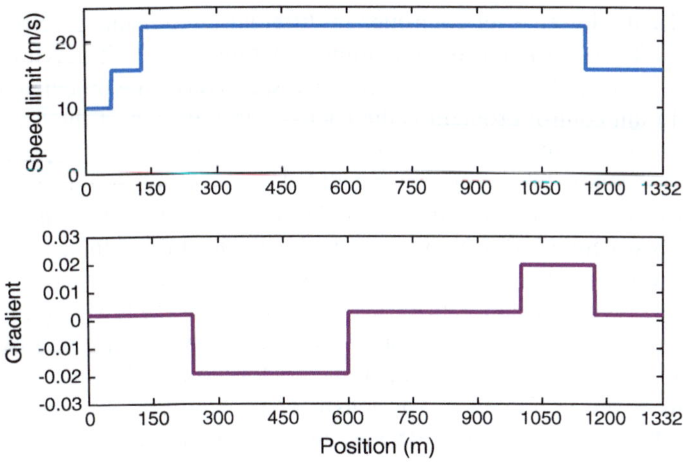

Fig. 4.3 The speed limits and the grade profile between Songjiazhuang station and Xiaocun station

Table 4.1 Parameters of the train and line path

Property	Symbol	Value
Train mass (kg)	m	2.780×10^5
Resistance (N/kg)	R_b	$1.421 \times 10^{-2} + 1.040 \times 10^{-4}v^2$
Mass factor (–)	ρ	1.06
Maximum velocity (m/s)	V_{max}	22.22
Section length (m)	s_{end}	1.332×10^3
Minimum kinetic energy (J)	E_{min}	0.1
Maximum traction force (N)	u_{max}	2.224×10^5
Maximum braking force (normal) (N)	u_{min}	-2.224×10^5

simultaneous variant. For the sake of simplicity, we only consider two[2] stations in the Yizhuang subway line: Songjiazhuang station and Xiaocun station. The track length between these two stations is 1332 m and the speed limits and grade profile are shown in Fig. 4.3. The parameters of the train and the line section are listed in Table 4.1. According to the assumptions made in Sect. 3.3, the unit kinetic energy should be larger than some positive threshold E_{min}. In this test case, the minimum kinetic energy is chosen as 0.1 J. The maximum traction force of the train is a nonlinear function of the train's velocity and the maximum value of this function is 3.150 $\times 10^5$ N. Moreover, the maximum acceleration and maximum deceleration used for trajectory planning along the line are chosen to be 0.8 m/s^2 and -0.8 m/s^2 respectively in order to make sure that the planned trajectory can be followed by the train

[2]Note however that the MILP approaches and the pseudospectral methods can also be applied if more than 2 stations are considered.

controlled by the lower level controller in different circumstances. Since the train mass here is 2.780×10^5 kg, the maximum traction force and maximum braking force are 2.224×10^5 N and -2.224×10^5 N, respectively. The objective function of the optimal train control problem is the energy consumption of the train operation without consideration of regenerative braking (cf. (4.21)) in this chapter.

In this case study, two trains are scheduled to run from Songjiazhuang station with a headway of 75 s to Xiaocun station. We consider two cases: the FBS system and the MBS system. Moreover, we assume that the leading train has a malfunction during the whole simulation and as a consequence its maximum speed is reduced to 40 km/h, i.e., 11.1 m/s. In addition, the leading train and the following train will arrive at different platforms in Xiaoxun station and the following train will overtake the leading train at Xiaocun station. The parameters for the calculation of the minimum headway are given in Table 4.2. The length of the train is 90 m and the reaction time of the driver is 1 s. For the FBS system, we assume that there exist four fixed block sections between Songjiazhuang station and Xiaocun station and all fixed block sections are of equal length, i.e., 333 m. The minimum headway of the FBS system can be calculated according to (4.3) and it is equal to 98.4 s. Based on the parameters of Table 4.2, the run-in/run-out time $t_{\text{in-out}}$ in (4.19) is equal to 44.6 s and the minimum headway of the MBS system equals 69.6 s. Note that the headway 75 s is smaller than the minimum headway of the FBS system (i.e., 98.4 s) and is larger than the minimum headway of the MBS system (i.e., 69.6 s).

In the MILP approach, the length Δs_k of the intervals $[s_k, s_{k+1}]$ for $k = 1, 2, \ldots, N$ depends on the speed limits, gradient profile, fixed block length, and so on. If the number of the space intervals N is bigger, then the accuracy will be better but the computation time of the MILP approach will be longer. In this case study, the number of space intervals N is chosen as 20, 40, and 60, respectively. Moreover, the space intervals are taken to have equal length being 66.6 m, 33.3 m, and 22.2 m respectively for the three values of N. In addition, the nonlinear terms in the trajectory planning problem, such as the nonlinear terms in the differential equations of the train model, are approximated by piecewise affine functions (see Chap. 3 for more details). If we take piecewise affine approximations with more subfunctions, the approximation accuracy will be better. Here, the performance of the piecewise

Table 4.2 Parameters for the calculation of the minimum headway [1]

Property	Symbol	Value
Train length (m)	L_t	90
Safety margin (m)	S_{SM}	30
Length of the secure section (m)	L_s	60
Initial acceleration (m/s^2)	a_{acc}^{L}	1
Braking deceleration (m/s^2)	a_{dec}^{F}	0.9
Braking reaction time (s)	t_r	1
Station dwell time (s)	t_d	25

affine functions with 2 subfunctions and with three subfunctions are compared. We use the CPLEX solver via the Tomlab[3] interface to MATLAB® for solving the MILP problems.

The pseudospectral method is a state-of-the-art method for solving optimal control problems (cf. Chap. 3, Appendix A, [12, 13]). The approximation error of the pseudospectral method can be reduced by taking more collocation points. The numbers of LGL points are taken as 20, 40, 80, and 120, respectively. There are several packages that implement the pseudospectral method (see Appendix A for detailed information). One of them is PROPT, which supports the description of the differential algebraic equations and can call many solvers, such as MINOS and SNOPT, to solve the resulting nonlinear programming problem. We use PROPT solver through the Tomlab interface to MATLAB in our case study, where SNOPT is adopted to solve the resulting nonlinear programming problem. Note that both SNOPT and PROPT are implemented in object code.

4.6.2 Results for the Fixed Block Signaling System

Table 4.3 shows the performance of the MILP approaches and the pseudospectral methods for the trajectory planning of two trains in the FBS system. The performance mentioned here, such as the energy consumption and the end time violation, is computed by applying the obtained optimal control inputs into the nonlinear train model (3.1). The total energy consumption is the sum of the energy consumption of the leading train and the following train. The end time violation is the sum of the absolute values of the differences between the real running times and the planned running times for the leading train and the following train. The energy consumption for each train is influenced by the difference between the real running time and the planned running time. If the real running time is larger than the planned running time, e.g., 105 s for the following train, then the energy consumption usually becomes less since a train could run with a smaller average speed. For the greedy approach, the CPU time is the sum of the time used to solve the optimal control problem for the leading train and the optimal control problem for the following train on a 1.8 GHz Intel Core 2 Duo CPU running a 64-bit Linux operating system. In the simultaneous approach, the total CPU time is equal to the time spent by solving the optimal control problem since the optimal control inputs for the leading train and the following train are computed simultaneously.

Furthermore, Figs. 4.4, 4.5, 4.6 and 4.7 show the speed-position trajectories and the optimal control inputs obtained by the MILP approaches and the pseudospectral methods for the FBS system, where both the number of space intervals for the MILP approaches and the number of the collocation points for the pseudospectral methods are taken as 40. In addition, the nonlinear terms in the train model are approximated

[3]Tomlab website: http://tomopt.com.

Table 4.3 Performance comparison of the greedy and simultaneous approach using the MILP approach and the pseudospectral method for the FRS system

Approach	Method	Variant[a]	Problem size[b]	Total energy consumption (MJ)	Total end time violation	Total CPU time (s)
Greedy	MILP	2/20	L: 99/40/462 F: 199/140/1116	114.483	3.165	1.340
		2/40	L: 199/80/922 F: 399/280/2326	113.772	1.948	14.232
		2/60	L: 299/120/1382 F: 599/420/3486	110.597	1.476	24.303
		3/20	L: 179/120/962 F: 279/220/1686	114.794	2.821	6.865
		3/40	L: 359/240/1542 F: 559/440/3366	112.578	2.212	96.557
		3/60	L: 539/360/2702 F: 839/660/5046	110.028	1.410	229.723
	Pseudo-spectral	20	L: 60/0/1107 F: 60/0/1307	112.587	2.524	231.833
		40	L: 120/0/1207 F: 120/0/1607	110.041	1.429	1381.708
		80	L: 240/0/1407 F: 240/0/2207	109.478	0.761	1935.090
		120	L: 360/0/1607 F: 360/0/2807	109.173	0.449	3588.102
Simultaneous	MILP	2/20	358/180/1837	109.651	4.072	2.757
		2/40	718/360/3697	108.062	2.906	78.401
		2/60	1078/540/5557	106.317	1.652	204.380
		3/20	518/340/2877	108.436	3.190	24.134
		3/40	1038/680/5777	106.578	1.731	184.755
		3/60	1558/1020/8677	106.191	1.079	349.312
	Pseudo-spectral	20	120/0/2414	109.528	3.576	445.511
		40	240/0/2814	106.442	1.679	1521.302
		80	480/0/3614	105.917	0.784	3005.713
		120	720/0/4414	105.683	0.527	4875.230

[a]For the greedy and simultaneous MILP approaches, the n in the notation n/m is the number of subfunctions used in the piecewise affine approximations of the nonlinear terms in the differential equations of the nonlinear train model and m is the number of the space intervals. For the greedy and simultaneous pseudospectral methods, the n indicates the number of collocation points used

[b]For all four approaches, the number n_1 in the notation $n_1/n_2/n_3$ is the number of real-valued variables, n_2 is the number of integer-valued variables, and n_3 is the number of constraints. Note that for the greedy and simultaneous MILP approach, all the constraints are linear. Furthermore, in the greedy MILP and greedy pseudospectral methods, two subproblems are solved. One is for the trajectory planning problem of the leading train, the size of which is shown as L: $n_1/n_2/n_3$. The other is for the trajectory planning problem for the following train considering the constraints caused by the leading train, the size of which is shown as F: $n_1/n_2/n_3$. For the simultaneous MILP and simultaneous pseudospectral approaches, the trajectories of the leading train and the following train are obtained by solving a combined optimization problem

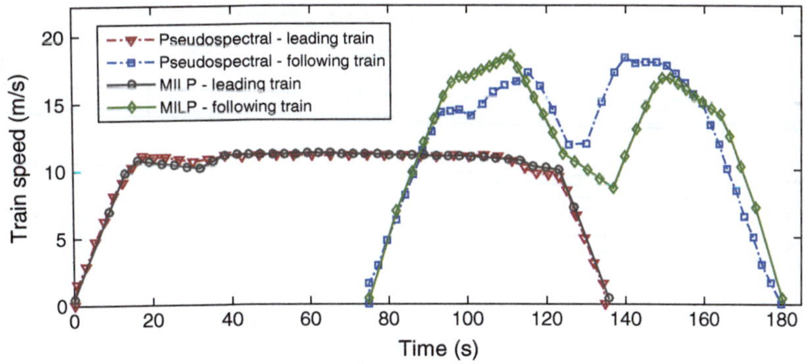

Fig. 4.4 Speed profiles of the two trains under the FBS system with headway 75 s using the greedy MILP and the pseudospectral approach

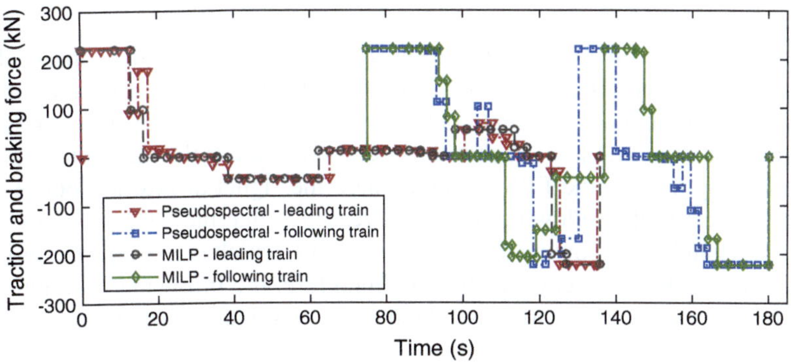

Fig. 4.5 Control inputs of the two trains under the FBS system with headway 75 s using the greedy MILP and the pseudospectral approach

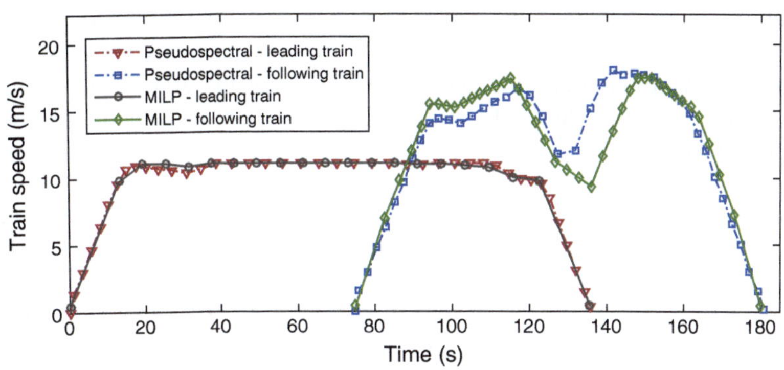

Fig. 4.6 Speed profiles of the two trains under the FBS system with headway 75 s using the simultaneous MILP and the pseudospectral approaches

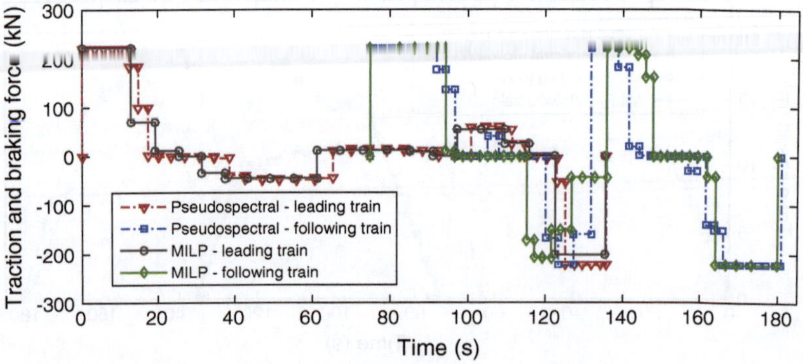

Fig. 4.7 Control inputs of the two trains under the FBS system with headway 75 s using the simultaneous MILP and the pseudospectral approaches

by piecewise affine functions with 3 affine subfunctions in the MILP approach. The speed-position trajectories of the leading train and the following train were generated by applying the obtained optimal control inputs to the nonlinear train model. It is observed from Figs. 4.4 and 4.6 that the operation of the following train is affected by the leading train in the FBS system, where the signal at position 666 m shows a yellow aspect to the following train. The reason is that the headway between the leading train and the following train is taken as 75 s, which is less than the minimum headway of the FBS system, i.e., 98.4 s. Therefore, the following train must slow down to satisfy the speed limit caused by the yellow signal aspect, which is 40 km/h, i.e., 11.1 m/s.

In the case study for the FBS system, the corresponding energy consumption, the end time violation, and the computation time of the greedy MILP approach are 112.578 MJ, 2.212 s, and 96.557 s, respectively. For the greedy pseudospectral method, the energy consumption, the end time violation, and the calculation time of both trains are 110.041 MJ, 1.429 s, and 1381.708 s, respectively. It can be seen that the energy consumption and the end time violation of the greedy MILP approach are slightly larger than those of the greedy pseudospectral method. However, the computation time of the greedy pseudospectral method is more than one order of magnitude longer than that of the greedy MILP approach.

The energy consumption, end time violation, and calculation time are 106.578 MJ, 1.731 s, and 184.755 s, respectively, using the simultaneous MILP approach. For the simultaneous pseudospectral method, they are 106.442 MJ, 1.679 s, and 1521.302 s, respectively. When compared with the greedy MILP approach, the energy consumption and the end time violation of the simultaneous MILP approach become smaller since the leading train's trajectory can also be optimized with respect to the following train. However, the computation time of the simultaneous MILP approach becomes longer than that of the greedy MILP approach because the size of the optimization problem in the simultaneous MILP approach is almost doubled. In addition, the simultaneous pseudospectral method yields a better performance, i.e., lower energy consumption and lower end time violation, when comparing with the greedy pseudospectral method.

It can be observed in Table 4.3 that the energy consumption and the end time violation of the greedy MILP approach are generally larger than that of the greedy pseudospectral method for the same number of discrete intervals and collocation points. However, the computation time of the greedy pseudospectral method is one to two orders of magnitude higher than that of the greedy MILP approach. Furthermore, the energy consumption and the end time violation become smaller if we take more space intervals for the greedy MILP approach and more collocation points for the greedy pseudospectral method. The results obtained for the greedy MILP approach and the greedy pseudospectral method also hold for the simultaneous MILP approach and the simultaneous pseudospectral method. Moreover, it can be observed that the energy consumption of the simultaneous MILP approach is less than that of the greedy MILP approach, while the computation times of the simultaneous MILP approach are higher than those of the greedy MILP approach. This also holds for the greedy and simultaneous pseudospectral method.

4.6.3 Results for the Moving Block Signaling System

The performance of the greedy and simultaneous approaches is given in Table 4.4 for the MBS system. Figures 4.8, 4.9, 4.10 and 4.11 show the speed-position trajectories and the optimal control inputs obtained by the greedy and the simultaneous approaches for the MBS system, where the set-up is the same as that stated in Sect. 4.6.2. In addition, the speed-position trajectories, the end time violation, etc. are obtained by applying the optimal control inputs to the nonlinear train model (3.1). As we can see from Figs. 4.8 and 4.10, the operation of the following train is not affected by the leading train since the scheduled headway (75 s) is larger than the minimum headway of the MBS system (69.6 s) (4.19).

In the case study for the MBS system, the energy consumption, the end time violation, and the computation time for the greedy MILP approach are 67.894 MJ, 1.682 s, and 75.975 s, respectively. In addition, the energy consumption, the end time violation, and the computation time for the greedy pseudospectral method are 67.416 MJ, 1.657 s, and 729.243 s, respectively. The energy consumption and the end time violation of the greedy MILP approach are slightly larger than those of the greedy pseudospectral method. However, the computation time of the greedy pseudospectral method is around one order of magnitude larger than that of the greedy MILP approach. This also holds for the results obtained by the simultaneous MILP and pseudospectral approach. In general, the total energy consumption and the total end time violation decrease with the increase of the number of the space intervals in the MILP approach and the number of collocation points in the pseudospectral method. Nevertheless, the total CPU time increases quickly with respect to the number of the space intervals and collocation points. For the greedy and/or simultaneous MILP approach, if we take piecewise affine approximations with more subfunctions, the end

Table 4.4 Performance comparison of the greedy and simultaneous approach using the MILP and the pseudospectral method for the MBS system

Approach	Method	Variant	Problem size	Total energy consumption (MJ)	Total end time violation (s)	Total CPU time (s)
Greedy	MILP	2/20	L: 99/40/462 F: 99/40/632	69.157	3.445	1.413
		2/40	L: 199/80/922 F: 199/80/1292	68.341	2.468	12.025
		2/60	L: 299/120/1382 F: 299/120/1952	67.558	1.083	24.191
		3/20	L: 179/120/962 F: 179/120/1152	68.762	2.791	9.869
		3/40	L: 359/240/1542 F: 359/240/2332	67.894	1.682	75.975
		3/60	L: 539/360/2702 F: 539/360/3662	67.116	1.335	254.796
	Pseudo-spectral	20	L: 60/0/1107 F: 60/0/1157	68.876	3.941	89.708
		40	L: 120/0/1207 F: 120/0/1257	67.416	1.657	729.243
		80	L: 240/0/1407 F: 240/0/1457	67.032	0.746	1483.420
		120	L: 360/0/1607 F: 360/0/1657	66.850	0.287	4542.203
Simultaneous	MILP	2/20	198/80/1094	68.141	2.584	2.446
		2/40	398/160/2214	67.520	1.533	56.169
		2/60	598/240/3334	66.615	1.262	150.671
		3/20	358/240/2134	67.732	1.707	13.486
		3/40	718/480/4294	67.031	1.042	121.292
		3/60	1078/720/6454	66.886	0.852	420.365
	Pseudo-spectral	20	120/0/2264	68.621	3.613	161.148
		40	240/0/2464	67.502	1.371	1051.059
		80	480/0/2864	67.069	0.661	2984.234
		120	720/0/3264	66.256	0.313	6954.372

time violation would decrease. Furthermore, when compared with the greedy MILP (or pseudospectral) approach, the simultaneous MILP (or pseudospectral) approach has a better control performance in general but it is characterized by a much higher computational burden since the size of the optimization problem is almost doubled.

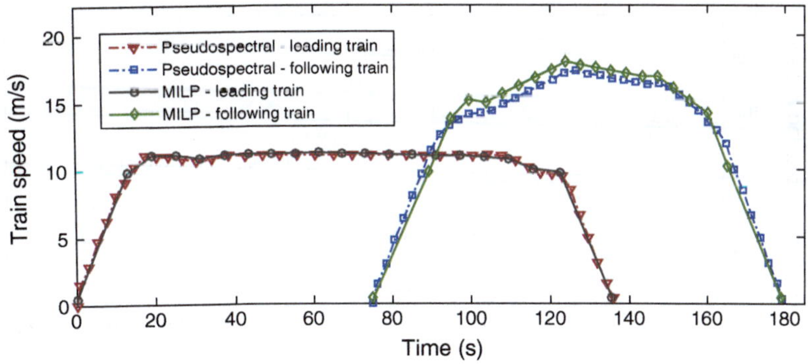

Fig. 4.8 Speed profiles of the two trains under the MBS system with headway 75 s using the greedy MILP and the pseudospectral approaches

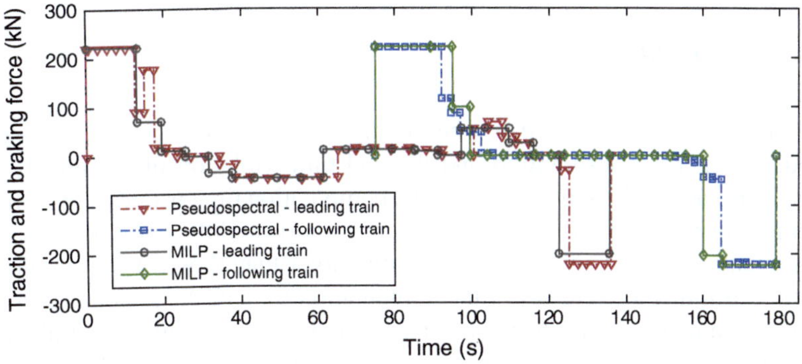

Fig. 4.9 Control inputs of two trains under the MBS system with headway 75 s using the greedy MILP and the pseudospectral approaches

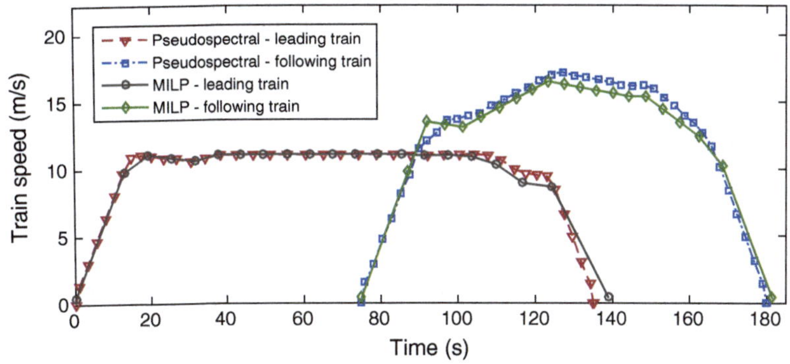

Fig. 4.10 Speed profile of the two trains under MBS system with headway 75 s using the simultaneous MILP and the pseudospectral approaches

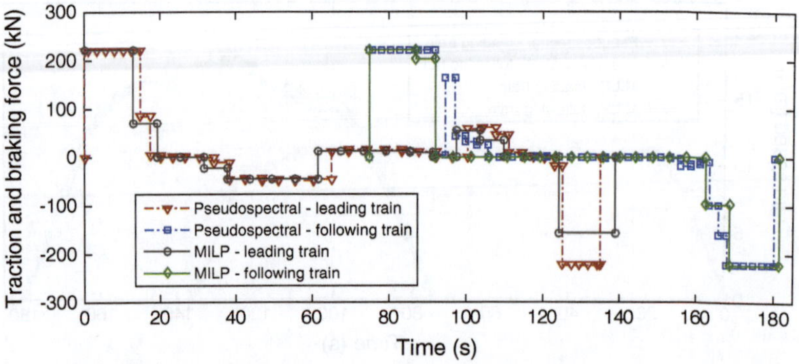

Fig. 4.11 Control input of the two trains under MBS system with headway 75 s using the simultaneous MILP and the pseudospectral approaches

4.6.4 Discussion

The simulation results show that when compared with the greedy pseudospectral method, the energy consumption and the end time violation of the greedy MILP approach are inconsiderably larger, but the computation time is one to two orders of magnitude shorter. Similarly, the energy consumption and the end time violation of the simultaneous MILP approach are slightly larger than those of the simultaneous pseudospectral method. However, the computation time of the simultaneous MILP approach is much smaller than that of the simultaneous pseudospectral method. Moreover, the energy consumption of the greedy MILP approach is larger than that of the simultaneous MILP approach, but the computation time of the simultaneous MILP approach is longer in general. Furthermore, the energy consumption of the simultaneous pseudospectral method is less than that of the greedy pseudospectral method, but it requires a much longer computation time.

4.7 Summary

In this chapter, we have proposed the greedy approach and the simultaneous approach to solve the optimal trajectory planning problem for multiple trains. In the greedy approach, the optimal trajectory planning problem of the leading train is solved at first. On the basis of the optimal control inputs of the leading train, the trajectory planning problem for the following train is then solved. For the simultaneous approach, the trajectories of the leading train and the following train are optimized at the same time. The constraints caused by the leading train in a FBS system and in a MBS system are included in the optimal trajectory planning problem for multiple trains. In the MILP approach, the nonlinear terms in the train model and constraints are approximated by piecewise affine functions. The optimal trajectory planning problem for multiple trains can then be recast as a MILP problem. The performance of the greedy and the simultaneous MILP approach has been compared with the greedy and the simultaneous pseudospectral method via a case study. The simulation results have illustrated that the MILP approach yields a similar control performance as the pseudospectral method but it requires a much shorter computation time.

A topic for future work will be an extensive comparison and assessment between the MILP approach, the pseudospectral method (also using other nonlinear programming subsolvers, e.g., MINOS and KNITRO), a dynamic programming algorithm [14], and other approaches and frameworks (such as AMPL, APMonitor, and ASCEND) described in the literature for various case studies and a wide range of scenarios.

References

1. Wang Y, De Schutter B, van den Boom T, Ning B (2014) Optimal trajectory planning for trains under fixed and moving signaling systems using mixed integer linear programming. Control Eng Pract 22:44–56
2. Wang Y, De Schutter B, van den Boom T, Ning B (2013) Optimal trajectory planning for trains under a moving block signaling system. In: Proceedings of the 12th European control conference, Zürich, Switzerland, pp 4556–4561
3. Wang Y, De Schutter B, van den Boom T, Ning B (2013) A suboptimal control scheme of multiple trains based on mode vector constraints. In: Proceedings of the 5th international seminar on railway operations modelling and analysis (RailCopenhagen), Copenhagen, Denmark
4. Takeuchi H, Goodman C, Sone S (2003) Moving block signaling dynamics: performance measures and re-starting queued electric trains. In: IEE Proceedings-electric power applications, vol 150, pp 483–492
5. Hill R, Bond L (1995) Modeling moving block railway signaling system using discrete-event simulations. In: Proceedings of the 1995 IEEE/ASME joint railroad conference, Baltimore, MD, USA, pp 105–111
6. Khmelnitsky E (2000) On an optimal control problem of train operation. IEEE Trans Autom Control 45:1257–1266
7. Liu R, Golovicher I (2003) Energy-efficient operation of rail vehicles. Transp Res Part A: Policy Pract 37:917–931
8. Camponogara E, Jia D, Krogh B, Talukdar S (2002) Distributed model predictive control. IEEE Control Syst Mag 22:44–52
9. Wang Y, De Schutter B, Ning B, Groot N, van den Boom T (2011) Optimal trajectory planning for trains using mixed integer linear programming. In: Proceedings of the 14th international IEEE conference on intelligent transportation systems (ITSC 2011), Washington, DC, USA, pp 1598–1604
10. Ingimundarson A, Ocampo-Martinez C, Bemporad A (2007) Suboptimal model predictive control of hybrid systems based on mode-switching constraints. In: Proceedings of the 46th IEEE conference on decision and control, New Orleans, LA, USA, pp 5264–5269
11. Ocampo-Martinez C, Ingimundarson A, Bemporad A, Puig V (2008) Suboptimal hybrid model predictive control: application to sewer networks. In: Proceedings of the 17th IFAC world congress, Seoul, Korea, pp 10021–10026
12. Elnagar G, Kazemi M, Razzaghi M (1995) The pseudospectral Legendre method for discretizing optimal control problems. IEEE Trans Autom Control 40:1793–1796

13. Gong Q, Ross I, Kang W, Fahroo F (2008) Connections between the covector mapping theorem and convergence of pseudospectral methods for optimal control. Comput Optim Appl 41:307–335
14. Hellström E, Slund JA, Nielsen L (2010) Design of an efficient algorithm for fuel-optimal look-ahead control. Control Eng Pract 18:1318–1327

Chapter 5
OD-Independent Train Scheduling for an Urban Rail Transit Line

Abstract In the previous two chapters, we have discussed trajectory planning for trains in a railway network based on given train schedules. In this chapter, the train scheduling problem based on origin–destination-independent (OD-independent) passenger demands for an urban rail transit line is considered with the aim of minimizing the total travel time of passengers and the energy consumption of trains. We propose a new iterative convex programming (ICP) approach to solve this train scheduling problem. Via a case study inspired by the Beijing Yizhuang line, the performance of the ICP approach is compared with other alternative approaches, such as nonlinear programming approaches, a mixed integer nonlinear programming (MINLP) approach, and a mixed integer linear programming (MILP) approach. The research discussed in this chapter is based on Wang et al. (Efficient real-time train scheduling for urban rail transit systems using iterative convex programming. IEEE Trans Intell Transp Syst 16:3337–3352, 2015) [1]; Wang et al. (Proceedings of the 1st IEEE international conference on intelligent rail transportation (2013 IEEE ICIRT), Beijing, China, 2013) [2]; Wang et al. (Proceedings of the 16th international IEEE conference on intelligent transportation systems (ITSC 2013), The Netherlands, The Hague, pp 1334–1339, 2013) [3].

5.1 Introduction

As has been pointed out in Chap. 2, the train scheduling of urban rail transit systems becomes more and more important for reducing the operation costs of railway operators and for guaranteeing passenger satisfaction. In the urban rail transit systems considered here, the lines are assumed to be separated from each other and each direction of the line has a separate rail track. Therefore, trains do not overtake each other. In addition, for urban rail transit systems with high frequencies, it is not a major issue to the passengers anymore whether or not the train schedule is cyclic since new trains arrive at a station every 2–5 min. In practice, rail transport operators therefore do not announce the train schedule to passengers but only provide some information, such as that a train will arrive within 2 min. Hence, rail transport operators can schedule trains in real time based on the current situation, such as the

Y. Wang et al., *Optimal Trajectory Planning and Train Scheduling for Urban Rail Transit Systems*, Advances in Industrial Control,
DOI 10.1007/978-3-319-30889-0_5

number of waiting passengers at stations, the passenger arrival rates, and the number and position of running trains [1–3].

In this chapter, we propose a real-time scheduling approach to schedule trains in a receding horizon way based on the OD-independent passenger demands for an urban rail transit line, where a predefined timetable or service headway is not needed. The train scheduling problem is essentially a multi-objective optimization problem because it should consider both the benefits of the rail operators and the passengers [4–7]. This chapter considers multi-objective optimization for the train scheduling problem, where the energy consumption of the trains and the total travel time of passengers are minimized. Since the train scheduling problem here is a non-smooth non-convex problem, we propose a new iterative convex programming (ICP) approach to solve the problem. The performance of the ICP approach is compared with a pattern search method, a mixed integer nonlinear programming (MINLP) approach, a mixed integer linear programming (MILP) approach, and a sequential quadratic programming (SQP) method.

The rest of this chapter is structured as follows. Section 5.2 formulates the evolution equations for the arrivals and departures of trains, the passenger demand characteristics, and the passenger and vehicle interaction. Section 5.3 describes the multi-objective cost function and the constraints of the real-time train scheduling problem. Section 5.4 presents several solution approaches for the train scheduling problem, in particular, the new ICP approach. In addition, we show that the problem with stop-skipping can be solved using an MINLP approach or an MILP approach. Section 5.5 further extends the train scheduling model to a model with stop-skipping at small stations. Section 5.6 compares the performance of the proposed solution approaches in Sect. 5.4 with a case study. Finally, conclusions and recommendations are provided in Sect. 5.7.

5.2 Model Formulation

This chapter considers one direction of an urban transit line consisting of J stations as shown in Fig. 5.1. Station 1 is the starting station and station J is the final station of each trip.

We make the following assumptions when formulating the real-time scheduling model:

A.1 Station j for $j \in \{2, 3, \ldots, J-1\}$ can only accommodate one train at a time and no passing can occur at any point in the line.

A.2 Passengers arrive at a constant rate[1] λ_j at station j.

A.3 The number of passengers alighting from trains at station j for $j \in \{1, 2, \ldots, J\}$ is a fixed proportion ρ_j of the number of passengers on board of trains when they arrive at station j.

[1] Here we use a deterministic model to describe the passenger arrival process.

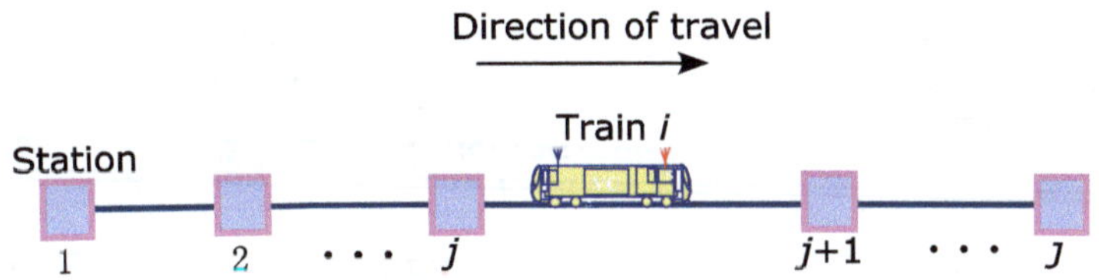

Fig. 5.1 Illustration of an urban rail transit line

A.4 The number of passengers waiting at a station and the number of passengers on board immediately after a train's departure are approximated by real numbers.
A.5 The operation of trains only consists of three phases: the acceleration phase, the speed holding phase, and the deceleration phase. Moreover, the acceleration and the deceleration are taken to be fixed constants.

Assumption A.1 generally holds for most urban transit systems, which are usually operated in first-in, first-out order from station 1 to J. Assumption A.2 is consistent with observed random passengers' arrivals for short headway (less than 10 min) services [8]. An estimate of these passenger arrival rates λ_j at stations can be obtained by analyzing historical data of the passenger flow. Assumption A.3 is made according to [9]. Similar as the passenger arrival rate, the passenger alighting proportion ρ_j can be estimated by analyzing historical data. For Assumption A.4, if the number of passengers is high, then the error made by this assumption is small. Furthermore, this assumption simplifies the optimization problem later on. In order to simplify the operation model for the trains, the detailed dynamics are not included in the model formulation, but only the three phases mentioned in Assumption A.5 are considered. However, once the running times between consecutive stations are fixed, a more accurate speed profile for the operation of trains can be calculated as a lower level control problem (see Chap. 3 for more information).

5.2.1 *Arrivals and Departures*

In the literature on train scheduling [7, 10–13], the operation of trains is usually described by the departure times, arrival times, running times, and dwell times. As has been discussed in Chap. 2, the operation of trains is controlled through a multilayer control framework. This chapter focuses on the train scheduling. In the scheduling layer, we use an online model-based approach; this means the model needs to be simulated repeatedly. In order to obtain a balanced trade-off between the accuracy and the computation speed, a macroscopic model is used. The detailed train dynamics, the position of block signals, the detection of trains, etc., can then be considered by the lower level control layer.

The departure time $d_{i,j}$ as shown in Fig. 5.2 of train i at station j is

$$d_{i,j} = a_{i,j} + \tau_{i,j}, \tag{5.1}$$

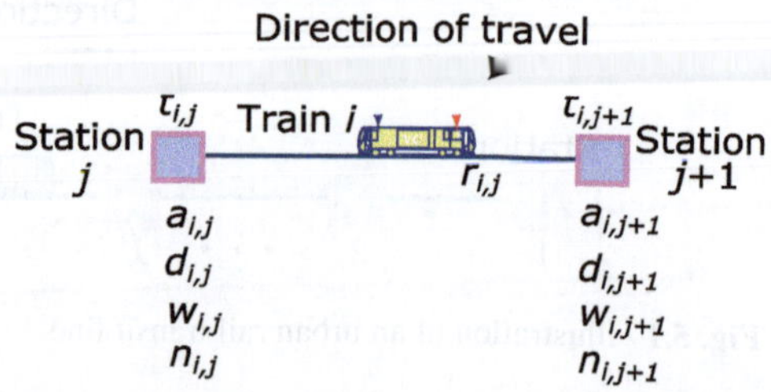

Fig. 5.2 The variables for the operation of trains and passenger characteristics

where $a_{i,j}$ and $\tau_{i,j}$ are the arrival time and the dwell time of train i at station j. In the literature, the dwell time is usually regarded as a constant. However, it is influenced by the number of passengers boarding and alighting from a train in practice. Therefore, we consider a variable dwell time, as will be explained in Sect. 5.2.3.

The track section between station j and station $j+1$ is denoted as segment j as illustrated in Figure 5.2. So the arrival time $a_{i,j+1}$ of train i at station $j+1$ equals the sum of the departure time $d_{i,j}$ at station j and the running time $r_{i,j}$ on segment j for train i:

$$a_{i,j+1} = d_{i,j} + r_{i,j}. \tag{5.2}$$

Remark 5.1 Max-plus algebra has been adopted for the train rescheduling and timetable analysis [14–16]. However, it is not suitable for the problem proposed here because the max operation will not be used since there is no fixed timetable yet. Moreover, it would result in a bilinear varying max-plus model since both running times and dwell times are variables.

According to Assumption A.5, we denote the acceleration and the deceleration as $a_{i,j}^{\text{acc}}$ and $a_{i,j}^{\text{dec}}$, respectively. If we define the speed in the speed holding phase as $v_{i,j}$, then the running distance of these three phases can be calculated as

$$s_{i,j}^{\text{acc}} = \frac{v_{i,j}^2}{2a_{i,j}^{\text{acc}}}, \quad s_{i,j}^{\text{dec}} = -\frac{v_{i,j}^2}{2a_{i,j}^{\text{dec}}}, \quad s_{i,j}^{\text{hold}} = s_j - s_{i,j}^{\text{acc}} - s_{i,j}^{\text{dec}},$$

where s_j is the length of segment j. Therefore, the running time of train i for segment j can be written as $r_{i,j} = t_{i,j}^{\text{acc}} + t_{i,j}^{\text{hold}} + t_{i,j}^{\text{dec}}$, which can then be recast as

$$r_{i,j} = \frac{s_j}{v_{i,j}} + \frac{v_{i,j}}{2a_{i,j}^{\text{acc}}} - \frac{v_{i,j}}{2a_{i,j}^{\text{dec}}}. \tag{5.3}$$

Note that the speed $v_{i,j}$ of the holding phase should satisfy

$$v_{i,j} \in [v_{i,j,\min}, v_{i,j,\max}], \tag{5.4}$$

where $v_{i,j,\min}$ and $v_{i,j,\max}$ are the minimal and maximal running speeds for the speed holding phase of train i at segment j, respectively. The maximum running speed is restricted by the train characteristics and the condition of the line. The minimum running speed is introduced to guarantee passenger satisfaction because the passengers may complain if trains run too slow.

Remark 5.2 The coasting phase of the operation of trains can be included as follows (at the cost of an increased number of variables and an increased computational complexity). In the coasting phase, both the traction force and braking force are equal to zero, where the train speed slows down because of the resistance. With the model given above, the entering speed (i.e., the holding speed) of the coasting phase is $v_{i,j}$. We denote the speed at the end of the coasting phase as $v'_{i,j}$. Note that the resistance varies with the train speed. Here, we approximate the resistance using the mean speed $v_{\mathrm{m},i,j} = (v_{i,j} + v'_{i,j})/2$ of the coasting phase:

$$R_i = m(k_{1i} + k_{2i}v_{\mathrm{m},i,j}) + k_{3i}v_{\mathrm{m},i,j}^2,$$

where m is the mass of the train itself and of the passengers on board of the train and k_{1i}, k_{2i}, and k_{3i} are the resistance coefficients of train i. In this way, the running distance and running time of the coasting phase can be computed. So we can include the coasting phase in the model formulation for the optimization of train schedule.

The minimum headway is defined as the minimum time interval between two successive trains such that they can enter and depart from a station safely [17]. A train cannot enter a station until a minimum headway h_0 after the departure of the preceding train, which can be written as

$$a_{i,j} - d_{i-1,j} \geq h_0. \tag{5.5}$$

Moreover, we select an order in which the trains run such that vehicle $i-1$ always precedes train i for $i \in \{1, 2, 3, \ldots, I\}$ with I the total number of trains.

5.2.2 *Passenger Demand Characteristics*

The number of passengers still remaining at the station immediately after the departure of train $i-1$ at station j is denoted as $w_{i-1,j}$. The number of passengers who want to get on train i at station j can then be formulated as

$$w_{i,j}^{\mathrm{wait}} = w_{i-1,j} + \lambda_j(d_{i,j} - d_{i-1,j}), \tag{5.6}$$

where $\lambda_j(d_{i,j} - d_{i-1,j})$ is the number of the passengers who arrived between the departure of train $i-1$ and the departure of train i.

By defining the number of passengers on train i immediately after its departure at station $j-1$ as $n_{i,j-1}$, the remaining capacity of train i at station j immediately after the alighting process of the passengers is

$$n_{i,j}^{\mathrm{remain}} = C_{i,\max} - n_{i,j-1}(1 - \rho_j), \tag{5.7}$$

where $C_{i,\max}$ is the effective maximal capacity of train i, and $n_{i,j-1}(1-\rho_j)$ is the number of passengers remaining on train i immediately after all the passengers who wanted to leave the train have gotten off. It is noted that the effective maximal capacity can be estimated on the basis of the data from the daily operations, where the distribution of onboard passengers and the effect of the distribution of waiting passengers on the platforms, etc., can be taken into account.

The number of passengers boarding train i at station j equals the minimum of the remaining capacity and the number of waiting passengers, i.e.,

$$n_{i,j}^{\text{board}} = \min\left(n_{i,j}^{\text{remain}}, w_{i,j}^{\text{wait}}\right). \tag{5.8}$$

The number of passengers at station j immediately after the departure of train i, i.e., the passengers who cannot get on train i, is then given by

$$w_{i,j} = w_{i,j}^{\text{wait}} - n_{i,j}^{\text{board}}. \tag{5.9}$$

The number of passengers on train i when it departs from station j is equal to the sum of the passengers arriving but not getting off at station j and the passengers boarding on train i at station j, which can be formulated as

$$n_{i,j} = n_{i,j-1}(1-\rho_j) + n_{i,j}^{\text{board}}. \tag{5.10}$$

5.2.3 Passenger and Vehicle Interaction

The dwell time is influenced by the number of alighting and boarding passengers, the number on waiting passengers at platform, and the distribution of passengers on board of a train and of passengers waiting at platform, etc. Based on [18], the minimum dwell time can be described by a linear model

$$\tau_{i,j,\min} = \alpha_{1,\text{d}} + \alpha_{2,\text{d}} n_{i,j-1}\rho_j + \alpha_{3,\text{d}} n_{i,j}^{\text{board}}, \tag{5.11}$$

or a nonlinear model

$$\tau_{i,j,\min} = \alpha_{1,\text{d}} + \alpha_{2,\text{d}} n_{i,j}^{\text{alight}} + \alpha_{3,\text{d}} n_{i,j}^{\text{board}} + \alpha_{4,\text{d}} \left(\frac{w_{i,j}^{\text{wait}}}{n^{\text{door}}}\right)^3 n_{i,j}^{\text{board}}, \tag{5.12}$$

where $\alpha_{1,\text{d}}$, $\alpha_{2,\text{d}}$, $\alpha_{3,\text{d}}$, and $\alpha_{4,\text{d}}$ are coefficients that can be estimated based on historical data, n^{door} is the number of doors of the train, and $w_{i,j}^{\text{wait}}/n^{\text{door}}$ is the average number of passengers waiting at each door. The dwell time $\tau_{i,j}$ should satisfy

$$\max\left(\tilde{\tau}_{\min}, \tau_{i,j,\min}\right) \le \tau_{i,j} \le \tau_{i,j,\max}, \tag{5.13}$$

where $\tilde{\tau}_{\min}$ is the minimum dwell time predefined by railway operator. The dwell time $\tau_{i,j}$ should be larger than the predefined minimum dwell time $\tilde{\tau}_{\min}$ and it should also be larger than the minimal dwell time $\tau_{i,j,\min}$ such that the passengers can get on and get off the train. In addition, $\tau_{i,j}$ should be less than a maximum dwell time $\tau_{i,j,\max}$ to ensure the passenger satisfaction.

5.3 Mathematical Formulation of the Train Scheduling Problem

On the basis of the model formulation in Sect. 5.2, we now formulate the real-time train scheduling problem. The total travel time of all passengers and the energy consumption caused by the operation of trains are minimized through a weighted sum strategy for the real-time train scheduling problem. The total travel time is the sum of the passenger waiting time and the passenger in-vehicle time. The passenger waiting time $t_{i,j}^{\text{wait}}$ at station j for train i includes the waiting time of both passengers left by the previous train $i-1$ and the newly arrived passengers, and it can be calculated by

$$t_{i,j}^{\text{wait}} = w_{i-1,j}(d_{i,j} - d_{i-1,j}) + \frac{1}{2}\lambda_j(d_{i,j} - d_{i-1,j})^2, \tag{5.14}$$

where the first term represents the waiting time of the passengers left by train $i-1$ at station j, and the second term represents the waiting time of randomly arriving passengers between the departures of train $i-1$ and train i. The passenger in-vehicle time for train i running from station j to $j+1$ consists of the running time for all passengers on train i after its departure from station j and the waiting time of the passengers who do not get off the train at station $j+1$, which can be formulated as

$$t_{i,j}^{\text{in-vehicle}} = n_{i,j}r_{i,j} + n_{i,j}(1 - \rho_{j+1})\tau_{i,j+1}. \tag{5.15}$$

The total travel time of passengers can be written as a weighted sum of the passenger waiting time and the passenger in-vehicle time, i.e.,

$$t_{\text{total}} = \sum_{i=1}^{I}\sum_{j=1}^{J-1}(\gamma_{\text{wait}}t_{i,j}^{\text{wait}} + t_{i,j}^{\text{in-vehicle}}). \tag{5.16}$$

Remark 5.3 Since the passengers usually feel that time goes slowly when they are waiting at the platform [19], a weight larger than one can be added to the passenger waiting time in the problem formulation.

In [20], the energy consumption is proportional to the resistance, which includes the rolling resistance, air resistance, and grade resistance. However, the computation of the energy consumption is more precise in this chapter because the operation model of a train considered here includes the acceleration phase, the speed holding phase, and the deceleration phase. The deceleration phase usually does not consume energy. In addition, if the electric motor on board of a train can be used as an electric generator (so-called regenerative braking), then the regenerative energy can be fed back into the power supply system. The energy consumptions for each phase are then calculated as follows:

- The energy consumption for the acceleration phase of train i on segment j is

$$E_{i,j}^{\text{acc}} = \int_0^{t_{i,j}^{\text{acc}}} \Big((m_{\text{e},i} + n_{i,j} m_{\text{p}}) \big(a_{i,j}^{\text{acc}} + k_{1i} + k_{2i} v(t) + g \sin(\theta_j) \big) + k_{3i} v^2(t) \Big) v(t) \text{d}t, \tag{5.17}$$

where $m_{\text{e},i}$ is the mass of train i itself, m_{p} is the mass of one passenger, $(m_{\text{e},i} + n_{i,j} m_{\text{p}})$ is the mass of train i and the passengers on board of train i at station j, k_{1i}, k_{2i}, and k_{3i} are the resistance coefficients of train i, $v(t)$ is equal to $a_{i,j}^{\text{acc}} t$, and θ_j is the grade profile of segment j.
- The energy consumption for the speed holding phase of train i on segment j is

$$E_{i,j}^{\text{hold}} = \int_{t_{i,j}^{\text{acc}}}^{t_{i,j}^{\text{acc}} + t_{i,j}^{\text{hold}}} \Big((m_{\text{e},i} + n_{i,j} m_{\text{p}})(k_{1i} + k_{2i} v_{i,j} + g \sin(\theta_j)) + k_{3i} v_{i,j}^2 \Big) v_{i,j} \text{d}t. \tag{5.18}$$

- The energy generated by the regenerative braking in the deceleration phase can be calculated as

$$E_{i,j}^{\text{dec}} = \beta_i^{\text{dec}} \int_{t_{i,j}^{\text{acc}} + t_{i,j}^{\text{hold}}}^{r_{i,j}} \Big((m_{\text{e},i} + n_{i,j} m_{\text{p}}) \big(|a_{i,j}^{\text{dec}}| - k_{1i} - k_{2i} v(t) - g \sin(\theta_j) \big) - k_{3i} v^2(t) \Big) v(t) \text{d}t, \tag{5.19}$$

where β_i^{dec} is the ratio for regenerative energy fed back to the power supply system since there is power loss in electric generator of train i.

The total energy consumption for all I trains running with J stations can then be formulated as

$$E_{\text{total}} = \sum_{i=1}^{I} \sum_{j=1}^{J} (E_{i,j}^{\text{acc}} + E_{i,j}^{\text{hold}} - E_{i,j}^{\text{dec}}). \tag{5.20}$$

In order to solve the multi-objective optimization of the train scheduling problem, we apply a weighted sum strategy. So the objective function can be formulated as

$$f_{\text{opt}} = \frac{E_{\text{total}}}{E_{\text{total,nom}}} + \gamma \frac{t_{\text{total}}}{t_{\text{total,nom}}}, \tag{5.21}$$

where γ is a nonnegative weight, and the normalization factors $E_{\text{total,nom}}$ and $t_{\text{total,nom}}$ are the nominal values of the total energy consumption and the total travel time of passengers, respectively. These nominal values can be determined by running trains using a feasible initial schedule.

The constraints of the real-time scheduling problem include the running time constraints, dwell time constraints, headway constraints, and capacity of trains, given as (5.1)–(5.13) in Sect. 5.2.

Remark 5.4 The train scheduling problem presented here can be solved in a rolling horizon way. A detailed description on how to do this for train scheduling is given in Sect. 7.5. See there for more details.

5.4 Solution Approaches

The resulting train scheduling problem with objective function (5.21) and constraints (5.1)–(5.13) is a non-smooth non-convex problem, where the non-smoothness is caused by the min function in (5.8), and the non-convexity is due to the nonlinear non-convex objective function and the non-convex set defined by constraints. Several alternative approaches are proposed in this section to solve the train scheduling problem. One is a gradient-free non-smooth optimization approach, e.g., pattern search. Another one is a gradient-based nonlinear programming approach, e.g., SQP. Furthermore, a general purpose nonlinear integer programming approach, e.g., branch-and-bound algorithm, is also used. By approximating the nonlinear objective function using piecewise affine functions, the train scheduling problem can be recast into an MILP problem. Furthermore, a new ICP approach is presented to solve the train scheduling problem.

5.4.1 *Gradient-Free Nonlinear Programming*

Nonlinear programming approaches can be classified into gradient-free approaches and gradient-based approaches. The gradient-free approaches do not explicitly require gradient and Hessian information but only require that the values of the objective function can be ranked. Moreover, gradient-free methods are suitable for non-smooth problems. Due to the min function, the real-time train scheduling problem is non-smooth. Therefore, our first choice is to use a gradient-free method. In particular, we propose to use the pattern search method, since it can handle optimization problems with nonlinear, linear, and bound constraints, and does not require objective functions to be differentiable or continuous. The pattern search method was first proposed by Hooke and Jeeves [21], and it has been proved successful in practice even for objective functions with many local minima, in particular when used in combination with a multi-start method [22], in order to improve the probability of obtaining a solution that is close to a globally optimal solution.

When solving the scheduling problem by the pattern search method, the variables are the departure times $d_{i,j}$, the running times $r_{i,j}$, and the dwell times $\tau_{i,j}$. The other variables, such as the number of passengers $w_{i,j}$ waiting at stations, the number of passengers $n_{i,j}$ on board of the trains, the passenger waiting times $t_{\mathrm{wait},i,j}$, and the passenger onboard times $t_{\text{in-vehicle},i,j}$, can be eliminated. The pattern search method has been implemented in the global optimization toolbox of MATLAB® [23].

5.4.2 Gradient-Based Nonlinear Programming

Gradient-based nonlinear programming methods depend on gradient or Hessian information. If this information is not available, it can be approximated numerically. We consider the gradient-based SQP algorithm here since it is widely used to solve nonlinear programming problems. A requirement for the SQP algorithm is that the objective function and the constraints of the nonlinear programming problem should be continuously differentiable [24]. In the SQP method, the nonlinear programming problem is recast as a sequence of quadratic programming problems, which can be solved easily and efficiently. The train scheduling problem considered in this chapter is non-differentiable due to the min function. Even though the SQP algorithm is a gradient-based method, we also apply it to our problem setting[2] (since it yields good solutions in the case study in practice in Sect. 5.6).

When solving the real-time scheduling problem using the SQP algorithm, the variables are the same as those in the pattern search method. The SQP algorithm has been implemented in many packages, such as SNOPT [25] and the optimization toolbox of MATLAB [26].

5.4.3 Mixed Integer Nonlinear Programming

In the MINLP approach, we introduce auxiliary binary variables and auxiliary real variables to deal with the non-smooth min function in (5.8). By introducing a binary variable $\delta_{i,j} \in \{0, 1\}$ and defining

$$\tilde{f}_{i,j} = w_{i-1,j} + \lambda_j(d_{i,j} - d_{i-1,j}) - [C_{i,\max} - n_{i,j-1}(1 - \rho_j)], \tag{5.22}$$

the following equivalence holds [27]:

$$[\tilde{f}_{i,j} \leq 0] \Leftrightarrow [\delta = 1], \tag{5.23}$$

which is true iff

$$\begin{cases} \tilde{f}_{i,j} \leq \tilde{M}_{i,j}(1 - \delta_{i,j}) \\ \tilde{f}_{i,j} \geq \varepsilon + (\tilde{m} - \varepsilon)\delta_{i,j} \end{cases}, \tag{5.24}$$

[2]When the SQP algorithm is applied to a nonlinear programming problem with a non-differentiable objective function, it might get stuck in a local solution. In the nonlinear programming problem proposed in this chapter, the minimum value of the objective function is usually not obtained at the points where the objective function is non-differentiable, so the SQP algorithm will jump over these points. Therefore, the SQP approach with multiple initial points works well in this case.

where ε is a small positive number that is introduced to transform a strict equality into a non-strict inequality, and $\tilde{M}_{i,j}$ and $\tilde{m}_{i,j}$ are the maximum value and the minimum value of $\tilde{f}_{i,j}$, respectively. Equation (5.8) can now be rewritten as

$$n_{i,j}^{\text{board}} = \delta_{i,j}[w_{i-1,j} + \lambda_j(d_{i,j} - d_{i-1,j})] + (1 - \delta_{i,j})[C_{i,\max} - n_{i,j-1}(1 - \rho_j)]. \tag{5.25}$$

It should be noted that in (5.25) there are four nonlinear terms (i.e., $\delta_{i,j}w_{i-1,j}, \delta_{i,j}d_{i,j}$, $\delta_{i,j}d_{i-1,j}$, $\delta_{i,j}n_{i,j-1}$), which are the products of the binary variable $\delta_{i,j}$ and real variables. Four auxiliary real variables can then be introduced to transform these four nonlinear terms into linear terms with linear constraints. The detailed information about this transformation is given in Sect. 3.4.2.

The variables of the resulting MINLP problem involve the departure times $d_{i,j}$, the running times $r_{i,j}$, and the dwell times $\tau_{i,j}$, the passengers waiting at stations $w_{i,j}$, the passengers on board of the trains $n_{i,j}$, the binary variables $\delta_{i,j}$, and the auxiliary variables. The other variables like the passenger waiting times $t_{\text{wait},i,j}$ and the passenger onboard times $t_{\text{in-vehicle},i,j}$ can be eliminated. The MINLP problem can be solved using branch-and-bound method, such as the MINLP BB solver[3] and solving constraint integer programs (SCIP) [29]. Because the computation time of the MINLP BB solver are too long, so we propose a bi-level optimization method to solve this MINLP problem. This method consists of two levels of optimization. The high-level optimization optimizes the binary variables, where a brute force approach can be used to find all the combinations for the binary variables when the size of the problem is small. Alternatively, integer programming approaches, such as branch-and-bound, cutting planes, and heuristics, can be applied in the high-level optimization. For each combination of binary variables, the nonlinear optimization problem in the lower level is now a smooth non-convex optimization problem with real-valued variables only, which can be solved by gradient-based approaches, such as an interior point algorithm. See Sect. 6.4.1 for more information on the bi-level approach.

5.4.4 Mixed Integer Linear Programming

In Chap. 3, we have shown that the MILP approach can be very efficient for train trajectory planning problems. As a result, we also apply the MILP approach to the real-time train scheduling problem. In this approach, we approximate the nonlinear terms in the objective function by piecewise affine approximations and then transform the non-smooth non-convex problem into an MILP problem. The MILP approach deals with the min function of (5.8) in the same way as the MINLP approach. In the MINLP problem in Sect. 5.4.3, the constraints are linear, but the objective function is nonlinear and non-convex. Therefore, in order to solve the real-time rescheduling problem as an MILP problem, we need to approximate the nonlinear terms, such as

[3]For more details about the MINLP BB solver, see [28].

$w_{i-1,j}d_{i,j}$, $n_{i,j}r_{i,j}$, and $n_{i,j}\tau_{i,j}$, as piecewise affine functions. These nonlinear terms are products of two real valued variables. Here, we use the general form xy to denote such a nonlinear term, which can be rewritten as [30, 31]

$$xy = \frac{1}{4}(x+y)^2 - \frac{1}{4}(x-y)^2. \tag{5.26}$$

Define $\phi = x+y$ and $\xi = x-y$. Then we have $xy = \frac{1}{4}\phi^2 + \frac{1}{4}\xi^2$, where the quadratic terms ϕ^2 and ξ^2, or z^2 in general notation, can be approximated by a piecewise affine function of the following form:

$$f_{\text{PWA}}(z) = \begin{cases} \alpha_1 z + \beta_1 & \text{for} \quad z \le Z_1, \\ \alpha_2 z + \beta_2 & \text{for} \quad z > Z_1. \end{cases} \tag{5.27}$$

For each nonlinear term in the objective function, the values of α, β, and Z_1 are optimized based on least-squares optimization (see Sect. 3.4.2 for more details). By introducing the binary variables and auxiliary variables, the product xy can be recast as a linear expression with linear constraints (see Sect. 3.4.2 for more information).

The variables of the resulting MILP problem include the variables in the MINLP problem and the binary variables and the auxiliary variables introduced by the approximations of nonlinear terms in the objective function. The MILP problem can be solved by branch-and-bound algorithms implemented in several existing commercial and free solvers, such as CPLEX, Xpress-MP, or GLPK [32, 33].

5.4.5 A New Approach: Iterative Convex Programming

The non-differentiability of the train scheduling problem is introduced by the min function in (5.8). In addition, the non-convexity of the problem is caused by the variables $w_{i,j}$ and $n_{i,j}$. Therefore, we propose the ICP approach, where we use estimated values $\hat{w}_{i,j}$ and $\hat{n}_{i,j}$ for $w_{i,j}$ and $n_{i,j}$, respectively. This eliminates the min function and the nonlinear terms $w_{i,j}d_{i,j}$, $w_{i,j}d_{i-1,j}$, $n_{i,j}r_{i,j}$, and $n_{i,j}\tau_{i,j+1}$ in the objective function. Therefore, the resulting optimization problem is a smooth and convex problem, which can be solved for the global optimum using, e.g., interior point algorithms [34], which are implemented in the MATLAB software CVX for convex programming.[4] Based on the optimum of the convex problem, the new estimated values for $w_{i,j}$ and $n_{i,j}$ can be calculated. By solving the convex problems iteratively, we can get an approximation of the global optimum of the original non-smooth non-convex problem. The procedure of the ICP method is given in Algorithm 1.

[4]More information about MATLAB software CVX, see [35].

Algorithm 1 Iterative convex programming method

Input : a feasible initial solution of departure times, running times, and dwell times, i.e., $d_{i,j}(0)$, $r_{i,j}(0)$, and $\tau_{i,j}(0)$ for $i = 1, \dots, I$ and $j = 1, \dots, J$, convergence tolerance ς, maximum iteration number $p_{\max}$;

iteration number $p \leftarrow 0$;

compute initial estimates $\hat{w}_{i,j}(p)$ and $\hat{n}_{i,j}(p)$ using (5.9) and (5.10) according to $d_{i,j}(p), r_{i,j}(p)$, and $\tau_{i,j}(p)$;

calculate the initial objective $f_{\text{opt}}(p)$ by (5.21);

Repeat

    $p = p + 1$;

    substitute the estimates $\hat{w}_{i,j}(p-1)$ and $\hat{n}_{i,j}(p-1)$ into the original problem to get a convex problem;

    obtain optimal departure time $d^*_{i,j}(p)$, running time $r^*_{i,j}(p)$, and dwell time $\tau^*_{i,j}(p)$ by solving the resulting convex problem;

    compute $\hat{w}_{i,j}(p)$ and $\hat{n}_{i,j}(p)$ using (5.9) and (5.10) based on $d^*_{i,j}(p)$, $r^*_{i,j}(p)$, and $\tau^*_{i,j}(p)$;

    calculate the objective $f_{\text{opt}}(p)$ using (5.21);

Until $p = p_{\max}$ or $|f_{\text{opt}}(p) - f_{\text{opt}}(p-1)| \leq \varsigma$

$p^* = \arg\min_p f_{\text{opt}}(p)$;

Output : $d^*_{i,j}(p^*)$, $r^*_{i,j}(p^*)$, $\tau^*_{i,j}(p^*)$, $f_{\text{opt}}(p^*)$.

For the ICP approach, the variables of the real-time scheduling problem are the departure times $d_{i,j}$, the running times $r_{i,j}$, and the dwell times $\tau_{i,j}$. The number of passengers $w_{i,j}$ waiting at stations and the number of passengers $n_{i,j}$ on board of the trains are estimated by $\hat{w}_{i,j}$ and $\hat{n}_{i,j}$, respectively. The other variables, such as the passenger waiting time $t_{\text{wait},i,j}$ and the passenger onboard times $t_{\text{in}-\text{vehicle},i,j}$, are functions of the decision variables and hence are not explicitly represented in the solution process. The solution obtained by the ICP approach is not necessarily the global minimum of the formulated scheduling problem because the ICP approach deals with a sequence of convex approximations of the formulated nonlinear non-convex problem. For the ICP approach, we should in general employ multiple starting points. However, for the case study in Sect. 5.6, we found that one random feasible initial point yields comparable results with respect to other alternative approaches.

5.5 Extension: Stop-Skipping at Small Stations

In order to reduce the passenger travel time and energy consumption further, a stop-skipping scheme can be adopted, which has already been applied in practice, such as the SEPTA line in Philadelphia and the urban rail transit system in Santiago, Chile (see Chap. 2 for more detailed information). The stop-skipping strategy is beneficial for both the rail operator (e.g., less energy consumption and higher operation speed) and the passengers (e.g., shorter travel time and lower train occupation). With the help of the information provided via personal digital assistant (PDA) devices and the real-time displays and announcements at stations, we assume that passengers can obtain enough information and board the right train. However, the passengers at the skipped stations experience longer waiting time and thus a longer total travel time. Therefore, the skipping of trains at stations should be carefully coordinated. For example, additional constraints can be considered when scheduling the trains, such as: two successive trains should not skip the same station. In this way, the waiting time of passengers can be limited to an acceptable value.

In practice, if the passenger alighting proportion and the passenger arrival rate are high at a station, a train will not skip that station. Therefore, we assume here that trains may only skip small stations with low passenger alighting proportion ρ_j and that the effect of stop-skipping on passengers is then negligible for these small stations. Hence we define a skipping set $S_{\text{skip}} = \{(i, j)|\text{train } i \text{ may potentially skip station } j\}$.

We introduce a binary variable $y_{i,j}$ to indicate whether train i stops at station j or not:

$$y_{i,j} = \begin{cases} 1 \text{ if train } i \text{ will stop at station } j, \\ 0 \text{ if train } i \text{ will skip station } j. \end{cases} \tag{5.28}$$

For the sake of simplicity of the expression, we consider here the variables $y_{i,j}$ to be defined for pairs $(i, j) \in \{1, 2, \ldots, I\} \times \{1, 2, \ldots, J\}$. However, actually $y_{i,j}$ is only a free variable if $(i, j) \in S_{\text{skip}}$ and otherwise $y_{i,j} = 1$ by definition. Hence, instead of (5.1) we get

$$d_{i,j} = a_{i,j} + y_{i,j}\tau_{i,j}. \tag{5.29}$$

Since train i may skip station j or station $j + 1$, the running distance of the speed holding phase is then rewritten as

$$s_{i,j}^{\text{hold}} = s_j - y_{i,j}\frac{v_{i,j}^2}{2a_{i,j}^{\text{acc}}} + y_{i,j+1}\frac{v_{i,j}^2}{2a_{i,j}^{\text{dec}}}, \tag{5.30}$$

which means that if train i skips station j, then train i will run with the holding speed $v_{i,j}$ in the running distance of the acceleration phase. Similarly, if train i skips station $j + 1$, train i will run with the holding speed $v_{i,j}$ in the distance of the deceleration phase. Note that we have

$$(1 - y_{i,j+1})(v_{i,j+1} - v_{i,j}) = 0, \tag{5.31}$$

since when train i skips station $j + 1$, i.e., $y_{i,j+1} = 0$, the operation of the train between station j and station $j + 2$ only contains three phases.

The running time of train i for segment j can be written as

$$r_{i,j} = \frac{s_j}{v_{i,j}} + y_{i,j}\frac{v_{i,j}}{2a_{i,j}^{\text{acc}}} - y_{i,j+1}\frac{v_{i,j}}{2a_{i,j}^{\text{dec}}}. \tag{5.32}$$

The remaining capacity of train i at station j immediately after the passengers alight is reformulated as

$$n_{i,j}^{\text{remain}} = C_{i,\max} - n_{i,j-1}(1 - y_{i,j}\rho_j),$$

instead of (5.7). Instead of (5.8), the number of passengers boarding train i at station j can be calculated using

$$n_{i,j}^{\text{board}} = \min\left(n_{i,j}^{\text{remain}}, y_{i,j} w_{i,j}^{\text{wait}}\right), \tag{5.33}$$

where $y_{i,j} w_{i,j}^{\text{wait}}$ is the number of passengers who want to get on train i at station j. Furthermore, the number of passengers at station j immediately after the departure of train i can be computed by (5.9). Instead of (5.10), the number of passengers on train i when it departs from station j is now reformulated as

$$n_{i,j} = n_{i,j-1}(1 - y_{i,j}\rho_j) + n_{i,j}^{\text{board}}. \tag{5.34}$$

For the train scheduling problem with stop-skipping, the total energy consumption should be calculated as

$$E_{\text{total}} = \sum_{i=1}^{I}\sum_{j=1}^{J}(y_{i,j} E_{i,j}^{\text{acc}} + E_{i,j}^{\text{hold}} - y_{i,j+1} E_{i,j}^{\text{dec}}), \tag{5.35}$$

instead of (5.20). In addition, the passenger in-vehicle time for train i running from station j to $j+1$ is reformulated as

$$t_{\text{in-vehicle},i,j} = n_{i,j} r_{i,j} + y_{i,j+1} n_{i,j}(1 - \rho_{j+1})\tau_{i,j+1}, \tag{5.36}$$

instead of (5.15). The total travel time for the stop-skipping problem can then be calculated by (5.16).

The train scheduling problem with stop-skipping has objective function (5.21) and constraints (5.2), (5.4)–(5.9), and (5.29)–(5.14). This optimization problem is an MINLP problem, where the stoping variables $y_{i,j}$ are the binary variables and the objective function are nonlinear and non-convex. The MINLP approach and the MILP approach in Sects. 5.4.3 and 5.4.4 can be directly applied to solve the train scheduling problem with stop-skipping since they can easily deal with binary variables.

5.6 Case Study

5.6.1 Setup

In order to demonstrate the approaches proposed in Sect. 5.4 for the real-time train scheduling problem (with stop-skipping) and to compare their performance, the train characteristics and the line data of the Beijing Yizhuang subway line are used as a test case. The Beijing Yizhuang subway line has 14 stations as shown in Fig. 5.3, and the speed limit for the line is 80 km/h (i.e., 22.2 m/s). The detailed information

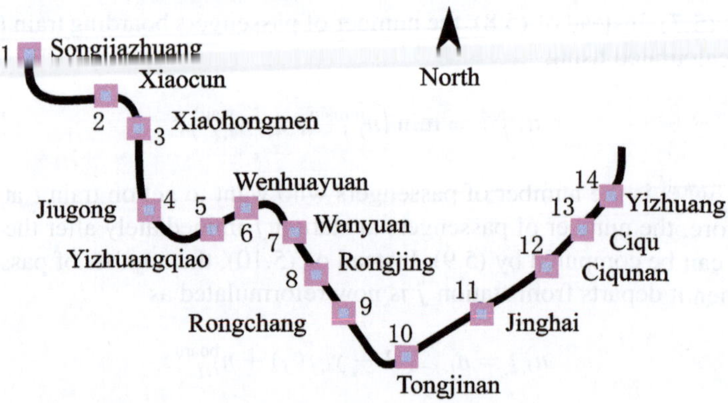

Fig. 5.3 The layout of the Beijing Yizhuang line

Table 5.1 Information of the Beijing Yizhuang line [1]

Station number	Distance to next station (m)	Minimum running time (s)	Maximum running time (s)	Passenger arrival rate (passenger/s)	Passenger alighting proportion (−)
1	1332	87.721	105.265	3	0
2	1286	85.651	102.781	0.5	0.05
3	2086	121.654	145.985	3	0.3
4	2265	129.710	155.652	4	0.38
5	2331	132.680	159.216	0.4	0.04
6	1354	88.711	106.453	4	0.32
7	1280	85.380	102.456	4	0.38
8	1544	97.260	116.712	3	0.7
9	992	72.420	86.904	3	0.6
10	1975	116.659	139.991	3	0.7
11	2369	134.391	161.269	3	0.7
12	1349	88.486	106.183	2	0.5
13	2610	145.237	174.284	2	0.5
14	–	0	0	1	–

about these 14 stations is given in Table 5.1. The minimum running time in Table 5.1 is calculated by taking a fixed acceleration of 0.8 m/s^2 and a fixed deceleration of −0.8 m/s^2. Furthermore, the speed $v_{i,j}$ of the holding phase in (5.3) is taken to be equal to the maximum speed 22.2 m/s for computing the minimum running time. It is assumed that the maximum running time is $r_{i,j,\max} = \zeta r_{i,j,\min}$, where ζ is larger than 1. We have chosen $\zeta = 1.2$ to ensure that the passengers do not complain that the train is too slow. According to the maximum running time, the minimal holding speed can be calculated.

Table 5.2 Parameters for trains and passengers [1]

Property	Symbol	Value
Train mass (kg)	$m_{\mathrm{e,i}}$	1.99×10^5
Mass of one passenger (kg)	m_{p}	60
Capacity of trains (passengers)	$C_{i,\max}$	1.468×10^3
Minimum dwell time (s)	$\tilde{\tau}_{\min}$	30
Maximum dwell time (s)	$\tau_{\max}$	150
Coefficients of the minimal dwell time		
(s)	$\alpha_{1,d}$	4.002
(s/passengers)	$\alpha_{2,d}$	0.047
(s/passengers)	$\alpha_{3,d}$	0.051
Coefficients of resistance		
(m/s^2)	k_{1i}	0.012
(s^{-1})	k_{2i}	5.049×10^{-4}
(m^{-1})	k_{3i}	2.053×10^{-5}

The mass of the train itself and the standard mass of one passenger are given in Table 5.2. In addition, we choose the linear model (5.11) for the minimum dwell time, the coefficients of which given in Table 5.2 are chosen according to [36]. The maximal dwell time is chosen as 150 s. The effective capacity of each train is 1468 passengers. A communication-based train control system (i.e., a moving block signaling scheme) is implemented in the Beijing Yizhuang line and the minimum headway h_0 between two successive trains is 90 s.

The initial state of the network at the start of the simulation is defined as follows: the train scheduling at the beginning of a day is considered. The schedule of train 0, i.e., the first train entering the urban rail transit line, is fixed. There are no passengers left by train 0 because not too many passengers wait for the first train in the morning. The schedule for the following trains should be optimized in this case study. Furthermore, in order to compare the performance of the schedules obtained by different approaches proposed in this chapter, a reference schedule with a fixed departure headway is defined, which is a state-of-the-art method used in practice, where the departure headway could change several times a day based on the peak hours and off-peak hours. For the scheduling period considered in this case study, the schedule of trains is taken the same as train 0 and the constant departure headway is taken as 210 s (i.e., the sum of the minimum headway and the maximum dwell time).

Two cases will be considered here:

- Case A: The train scheduling problem without stop-skipping.
- Case B: The train scheduling problem with stop-skipping.

For Case A, we have applied the approaches proposed in Sect. 5.4 to solve the train scheduling problem. For the pattern search method, we used the patternsearch function in the global optimization toolbox of MATLAB. The SNOPT solver implemented in the MATLAB Tomlab toolbox is adopted for the SQP algorithm to solve the nonlinear non-convex train scheduling problem. In the ICP approach, the resulting smooth and convex problem is also solved by the SNOPT. The MINLP BB solver is used for the MINLP approach. In addition, for the bi-level approach for MINLP problem, the low-level optimization problem is also solved using SNOPT and the ga function in the global optimization toolbox of MATLAB is applied for the high-level optimization. Furthermore, we use CPLEX, implemented through the cplex interface function of the MATLAB Tomlab toolbox as MILP solver.

Remark 5.5 SNOPT, MINLP BB, and CPLEX are implemented in object code while the patternsearch and ga functions are implemented in pure m files, i.e., they need to be interpreted and thus usually seem slower than the functions implemented in object code. So, in principle, the current computation time comparison is not fair. However, since we limited ourselves to the methods and functions available to us, we did our best to compare the results of these approaches in the fairest possible way. Note however that as explained above the computation time for the patternsearch and ga functions would be improved if they would also be implemented in object code.

In order to illustrate the performance of the solution approaches proposed for different-sized problems, we considered nine scenarios with different problem sizes as shown in Table 5.3, where the values of I and J are the number of trains and stations involved. For the scenarios with J less than 14, the passenger arrival rate λ_J and the passenger alighting rate ρ_J at station J is set to 0 and 1, respectively, because we assume that station J is the last station of the trip. The weight γ in the multi-objective function (5.21) is chosen as 1. In addition, the nominal values of the passenger travel time and the energy consumption shown in Table 5.3 are obtained using feasible train schedules, where the running times of trains are uniformly distributed random values in $[r_{\min}, 1.2 \cdot r_{\min}]$ and the dwell times are constants.

In Case B, we have applied the MINLP approach and the MILP approach to solve the train scheduling problem with stop-skipping based on Sect. 5.4. The resulting

Table 5.3 The nominal values of the energy consumption and the passenger travel time for all scenarios [1]

Scenario index	Values of I and J	Nominal passenger travel time (s)	Nominal energy consumption (J)
1	$I = 2, J = 3$	6.402×10^5	1.216×10^8
2	$I = 3, J = 4$	1.954×10^6	3.285×10^8
3	$I = 4, J = 5$	6.457×10^6	4.780×10^8
4	$I = 5, J = 6$	7.211×10^6	1.402×10^9
5	$I = 6, J = 7$	1.582×10^7	1.992×10^9
6	$I = 7, J = 8$	2.537×10^7	1.943×10^9
7	$I = 7, J = 10$	2.943×10^7	2.859×10^9
8	$I = 7, J = 12$	3.523×10^7	2.557×10^9
9	$I = 7, J = 14$	3.298×10^7	4.926×10^9

MINLP problem is solved using a direct MINLP approach, the brute force bi-level method, and the bi-level method with a genetic algorithm for the high-level binary optimization. The corresponding solvers and settings of these methods are chosen the same as those mentioned for Case A. According to the information given in Table 5.1, station 2 and station 5 are small stations since the passenger arrival rate and the passenger alighting proportion are smaller compared with other stations. In this case study, we allow trains to skip station 2 or/and station 5.

5.6.2 *Results and Discussion*

5.6.2.1 Results of Case A: The Train Scheduling Problem

The schedule of trains for scenario 5 obtained by solving the train scheduling problem for six trains (i.e., train $i \in \{1, 2, \ldots, 6\}$) and seven stations (i.e., station $j \in \{1, 2, \ldots, 7\}$) using the SQP approach is shown in Fig. 5.4. The running times, dwell times, and critical headways corresponding to Fig. 5.4 are shown in Table 5.4. The model formulation in this chapter allows the presence of waiting passengers at platforms at the beginning of the scheduling period and allows trains to be running somewhere on the transit line. As we can observe from Fig. 5.4, the departure headways between train 1, train 2, and train 3 at station 1 are larger than those between the later trains. This is because of the schedule of train 0, which stops at each station with a dwell time of 120 s. Therefore, in order to satisfy the headway constraints at all stations, the departure headway at the station 1 must be much larger than the minimum headway 90 s.

We applied a multi-start pattern search and a multi-start SQP here and we started the calculation with 10 feasible random initial points. As regards the ICP algorithm, we should also solve it using multiple initial points. However, we saw that the random feasible initial points yielded a comparable result with respect to each other. In addition, this result is also quite close to the solutions obtained by the other approaches.

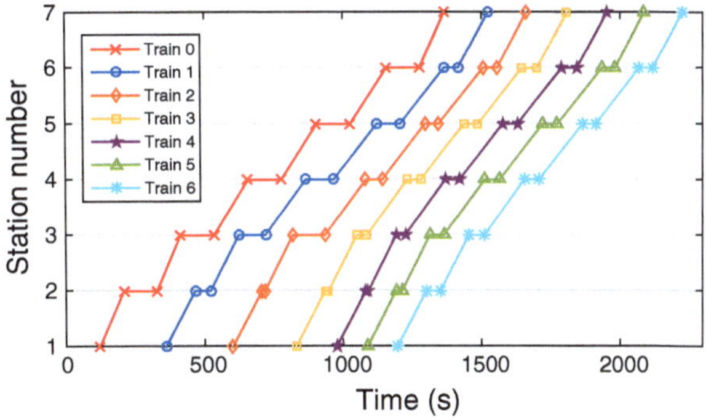

Fig. 5.4 The train schedule of scenario 5 with six trains and seven stations for Case A obtained by the sequential quadratic programming approach

Table 5.4 The total running times, dwell times, and critical minimum headways of each train for the train schedule of scenario 5 with six trains and seven stations for Case A obtained by the sequential quadratic programming approach

Train	Total running time (s)	Total dwell time (s)	Minimum headway (s)
0	646.128	720	–
1	775.354	538.132	90
2	775.354	438.250	90
3	775.354	343.417	90
4	775.354	256.282	90
5	775.354	245.964	90
6	775.354	272.063	90

Therefore, we use one feasible random initial point for the ICP approach in this case study. When we solve the MINLP problem using the bi-level optimization approach, the fmincon function in the lower optimization is executed for 10 feasible random initial points. For the MILP approach, only one feasible random initial point is needed to obtain the global minimum of the MILP problem.

The control performance f_{opt} and the computation time of these methods for the nine scenarios are illustrated in Figs. 5.5 and 5.6. The value of the objective function is influenced by the nominal values and weights in (5.21). Because we solve a minimization problem, a smaller value of the objective function means a better

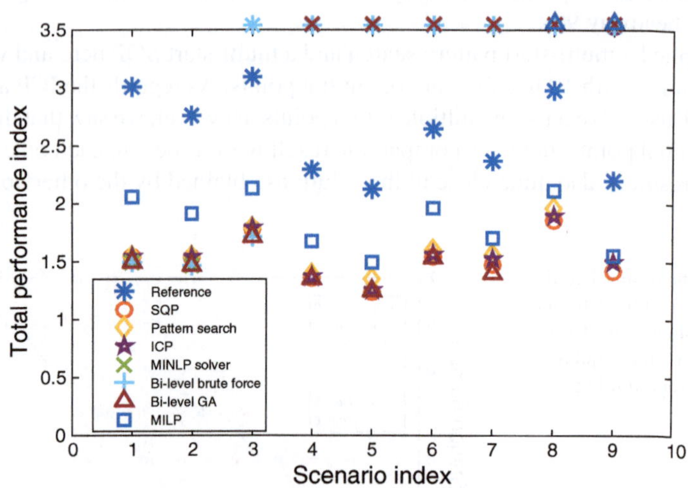

Fig. 5.5 Performance comparison of the solution approaches for the real-time train scheduling problem for Case A. In order to visualize the scenarios of which the computation cannot finish within 5 h, the performance index f_{opt} calculated using (5.21) of these scenarios is set larger than 3.5

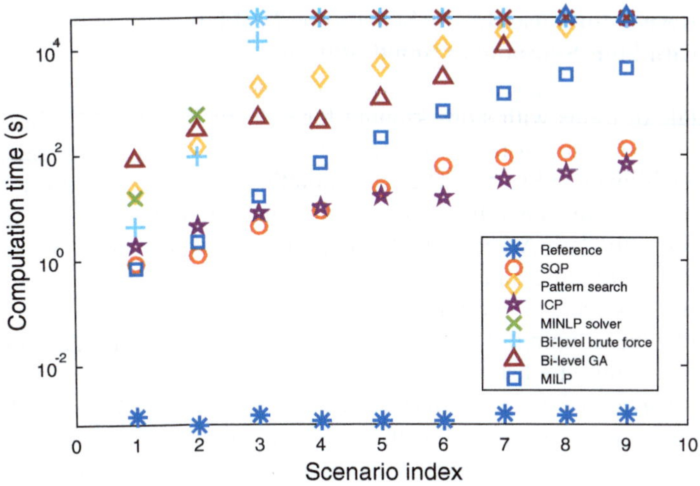

Fig. 5.6 The computation time of the solution approaches for the real-time train scheduling problem for Case A. In order to visualize the scenarios of which the computation cannot finish within 5 h, the computation time of these scenarios is set larger than 4×10^4 s

performance. It is noted that the control performance of the MILP approach is calculated using the original nonlinear objective function based on the resulting optimal results. We set the upper bound of the computation time as 5 h on a 64-bit Linux PC with 1.8 GHz Intel Core2 Duo CPU and 2 GB RAM. If the computation cannot be completed within 5 h, no results are reported, so we cannot determine the control performance index. In order to visualize the scenarios of which the computation cannot be finished within 5 h, we have set the total performance index of these scenarios larger than 3.5 as shown in Fig. 5.5 and set the computation time larger than 4×10^4 s as shown in Fig. 5.6. It can be noticed that only the performance indices of the SQP algorithm and the ICP algorithm are reported for scenario 9 in Fig. 5.5, and the calculation of other algorithms for scenario 9 cannot be finished within 5 h. It is observed that the reference schedule has the worst control performance but also the lowest computation time. In addition, the performance of the MILP approach is worse when compared with the other solution approaches that achieve similar control performance. The computation time of the ICP approach and the SQP approach is less than 15 s and 120 s for scenario 9, respectively. Furthermore, It can be observed that the function values of the ICP approach are mostly quite close to those of the solutions of the SQP approach, where for seven scenarios the relative performance difference is less than 5 % and for the other two scenarios the relative difference is around 10 %.

5.6.2.2 Results for Case B: The Scheduling Problem with Stop Skipping at Small Stations

The schedule of trains with stop-skipping for scenario 5 with six trains and seven stations obtained by the bi-level method with a genetic algorithm for the high-level optimization is given in Fig. 5.7. The corresponding running times, dwell times, and critical headways are shown in Table 5.5. Note that the normalize rate in (5.21) is the same for both Case A and Case B. It can be observed that train 3 and train 5 skip station 2 and train 2 and train 5 skip station 5 since trains are allowed to skip the small stations 2 and 5. Furthermore, the headways between trains at stations are influenced by the skipping of trains at station 2 and station 5 in order to satisfy the headway constraints at all stations.

The comparison of the performance index for Case A and Case B, obtained by the bi-level optimization approach with a genetic algorithm for the high-level integer optimization, is given in Table 5.6. It is observed that the performance indices of

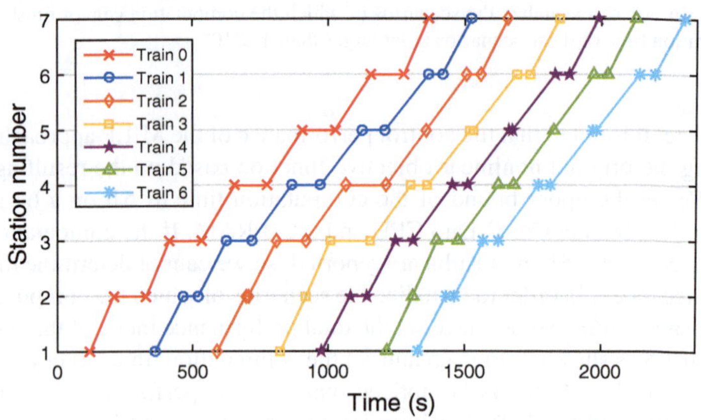

Fig. 5.7 The train schedule of scenario 5 with six trains and seven stations for Case B obtained by the bi-level optimization method with a genetic algorithm for the high-level optimization

Table 5.5 The total running times, dwell times, and critical minimum headways of each train for the train schedule of scenario 5 with 6 trains and 7 stations for Case B obtained by the bi-level optimization method with a genetic algorithm for the high-level optimization

Train	Total running time (s)	Total dwell time (s)	Minimum headway (s)
0	646.128	720	–
1	775.354	538.132	90
2	749.162	465.550	90
3	750.450	420.036	90
4	767.876	306.146	90
5	724.259	347.921	90
6	775.354	221.829	90

Table 5.6 Comparison of the total performance index for train scheduling with and without stop-skipping using the bi-level optimization with a genetic algorithm for the high-level integer optimization

	Scenario	1	2	3	4	5	6	7	8	9
Performance index (–)	Case A	1.500	1.473	1.713	1.360	1.246	1.532	1.407	–	–
	Case B	1.086	1.402	1.689	1.175	1.159	1.413	1.404	–	–
Total energy consumption (J)	Case A	1.328×10^8	3.379×10^8	6.589×10^8	1.102×10^9	1.602×10^9	2.165×10^9	2.673×10^9	–	–
	Case B	8.421×10^7	2.933×10^8	6.231×10^8	8.665×10^8	1.396×10^9	1.902×10^9	2.455×10^9	–	–
Total travel time (s)	Case A	2.610×10^5	8.683×10^5	2.161×10^6	4.137×10^6	6.993×10^6	1.060×10^7	1.388×10^7	–	–
	Case B	2.515×10^5	9.951×10^5	2.487×10^6	4.017×10^6	7.247×10^6	1.104×10^7	1.605×10^7	–	–

Case B, i.e., the train scheduling problem with stop-skipping, are better (i.e., a lower performance index means a better performance) than those of Case A, i.e., the train scheduling problem without stop-skipping, in general. The energy consumption is decreased since some trains do not need to decelerate and accelerate again at those small stations. Skipping some small stations may reduce the travel time of most passengers due to the zero dwell time at small stations and lower running times, however, it will increase the travel time of these passengers influenced by the skipping of trains. Therefore, the skipping of trains should be carefully coordinated.

Figures 5.8 and 5.9 show the performance index and the computation time of the MINLP approach and the MILP approach for the nine scenarios with stop-skipping.

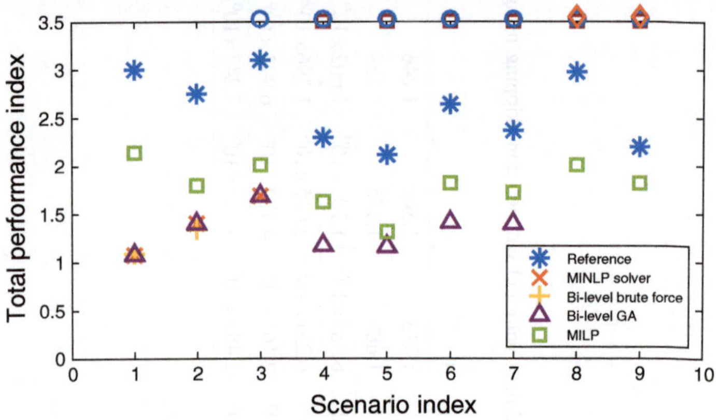

Fig. 5.8 Performance comparison of the solution approaches for the real-time train scheduling problem with stop-skipping for Case B. In order to visualize the scenarios of which the computation cannot finish within 5 h, the performance index f_{opt} calculated using (5.21) of these scenarios is set larger than 3.5

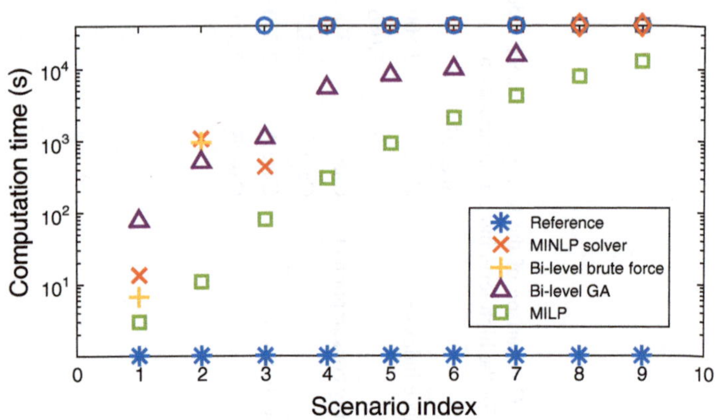

Fig. 5.9 The computation time of the solution approaches for the real-time train scheduling problem with stop-skipping for Case B. In order to visualize the scenarios of which the computation cannot finish within 5 h, the computation time of these scenarios is set larger than 4×10^4 s

The reference schedule is same as Case A and it has the lowest computation time and the worst performance. The performance index of the MILP approach is calculated by the original nonlinear objective function based on the obtained optimal results. Because of the errors introduced by the piecewise affine approximations of nonlinear terms in the objective function, the performance index of the MILP approach is much higher than other solution methods. The performance of the direct MINLP approach, the brute force bi-level optimization, and the bi-level optimization with a genetic algorithm for the real-time train scheduling problem with stop-skipping is similar to each other. However, we only report the results of scenario 1 and 2 for the brute force bi-level approach and the results of scenarios 1–3 for the direct MINLP approach since the computation for other scenarios could not be finished within 5 h. In addition, the bi-level optimization approach with a genetic algorithm for the high-level optimization could not finish the calculations within 5 h for scenarios 8 and 9. As a result, the performance indices of the MINLP approach and the MILP approach are set higher than 3.5 in Fig. 5.8. It is observed that the MILP approach needs less computation effort but at the cost of much less optimal performance. The bi-level optimization method with a genetic algorithm requires a longer computation time, but it yields a better performance.

5.6.2.3 Discussion

For the train scheduling problem in Case A, the performance of the results obtained by the ICP approach is close to the best results obtained by other alternative approaches. In addition, the computation time of the ICP approach is smaller than that of other alternative approaches except the reference schedule that has a bad performance. Therefore, for the given cases studies the ICP approach produces the best trade-off between performance and computational complexity.

According to the simulation results for Case B, the bi-level optimization with a genetic algorithm is recommended for solving the train scheduling problem with stop-skipping. Using parallel processing, the computation of this approach would still be tractable for small-sized problems (up to, say 20 stations and 10 trains). However, this approach is too slow for large-scale real-time applications. Therefore, new approaches need to be investigated in the future to solve the train scheduling problem with stop-skipping efficiently.

Remark 5.6 A stop-skipping strategy could lead to shorter circulation time of trains; so the turnaround operation at terminal stations may become critical for the operation of trains. However, the turnaround time can be reduced further with the increasing automation of the operation of trains, e.g., automatic turnaround of trains.

5.7 Summary

In this chapter, the train scheduling problem with OD-independent passenger demands for an urban rail transit line has been investigated. We have proposed a new ICP approach to solve this train scheduling problem. In addition, we have also considered other solution approaches, i.e., a gradient-free nonlinear programming approach (in particular pattern search method), a gradient-based nonlinear programming approach (in particular SQP), a MINLP approach (which includes three submethods, i.e., an MINLP approach, brute force bi-level optimization, and a bi-level optimization approach with a genetic algorithm), and a MILP approach. Furthermore, the train scheduling model with stop-skipping has been formulated by introducing binary variables to indicate whether a train stops at a station or not. The MINLP approach and the MILP approach are applied to solve this scheduling problem since they can handle integer variables.

The simulation results for train scheduling without stop-skipping have shown that the optimal solutions obtained by the ICP approach, the pattern search method, the SQP approach, and the MINLP approach are close to each other. However, the ICP approach can provide a better trade-off between performance and computational complexity. Furthermore, for the train scheduling problem with stop-skipping, the experiment results have shown that the control performance of the MILP approach is worse than that of the MINLP approach. Among the three submethods of the MINLP approach, the bi-level approach with a genetic algorithm offers the best trade-off between performance and computational efficiency.

In the future, one could investigate the effect of more detailed models (modeling the operation of trains in terminals, short turns, the stochastic variability in passenger flow, the distribution of onboard passengers and waiting passengers at platforms, the passengers appearing at platforms after the last train has passed, the passenger flows as described by origin–destination matrices) on the trade-off between performance and computational complexity. In addition, decomposition approaches can be developed to solve large-size instance of the train scheduling problem. Moreover, an extensive comparison and assessment of the approaches proposed in this chapter for a wide range of setups and scenarios could also be a topic for future work.

References

1. Wang Y, De Schutter B, van den Boom T, Ning B, Tang T (2015) Efficient real-time train scheduling for urban rail transit systems using iterative convex programming. IEEE Trans Intell Transp Syst 16:3337–3352
2. Wang Y, De Schutter B, van den Boom T, Ning B, Tang T (2013) Real-time scheduling in urban rail transit operations control. In: Proceedings of the 1st IEEE international conference on intelligent rail transportation (2013 IEEE ICIRT), Beijing, China. Paper 108
3. Wang Y, De Schutter B, van den Boom T, Ning B, Tang T (2013) Real-time scheduling for trains in urban rail transit systems using nonlinear optimization. In: Proceedings of the 16th international IEEE conference on intelligent transportation systems (ITSC 2013), The Netherlands, The Hague, pp 1334–1339

4. Vázquez-Abad F, Zubieta L (2005) Ghost simulation model for the optimization of an urban subway system. Discret Event Dyn Syst 15:207–235
5. Kwan C, Chang C (2005) Application of evolutionary algorithm on a transportation scheduling problem—the mass rapid transit. In: Proceedings of the IEEE congress on evolutionary computation, Edinburgh, UK, pp 987–994
6. Gassel C, Albrecht T (2009) The impact of request stops on railway operations. In: Proceedings of the 3rd international seminar on railway operations modeling and analysis, Zürich, Switzerland, pp 1–15
7. Higgins A, Kozan E, Ferreira L (1996) Optimal scheduling of trains on a single line track. Transp Res Part B 30:147–161
8. Okrent M (1974) Effects of transit service characteristics on passenger waiting times. M.S. thesis, Northwestern University, Evanston, USA
9. Simon J, Furth P (1985) Generating a bus route O-D matrix from on-off data. J Transp Eng 111:583–593
10. Hansen I, Pachl J (2008) Railway, timetable and traffic: analysis, modelling, simulation. Eurailpress, Hamburg, Germany
11. D'Ariano A, Pranzoand M, Hansen I (2007) Conflict resolution and train speed coordination for solving real-time timetable perturbations. IEEE Trans Intell Transp Syst 8:208–222
12. Elberlein X, Wilson N, Bernstein D (2001) The holding problem with real-time information available. Transp Sci 35:1–18
13. Li X, Wang D, Li K, Gao Z (2013) A green train scheduling model and fuzzy multi-objective optimization algorithm. Appl Math Model 37:617–629
14. Goverde R (2005) Punctuality of railway operations and timetable stability analysis. PhD thesis, Delft University of Technology, Delft, Netherlands
15. van den Boom T, Kersbergen B, De Schutter B (2012) Structured modeling, analysis, and control of complex railway operations. In: Proceedings of the 51st IEEE conference on decision and control, Maui, Hawaii, pp 7366–7371
16. Kersbergen B, van den Boom T, De Schutter B (2013) Reducing the time needed to solve the global rescheduling problem for railway networks. In: Proceedings of the 16th international IEEE conference on intelligent transportation systems (ITSC 2013), The Hague, The Netherlands, pp 791–796
17. Pachl J (2009) Railway operation and control, 2nd edn. Gorham Printing, Centralia
18. Lin T, Wilson N (1992) Dwell time relationships for light rail systems. Transp Res Rec 1361:287–295
19. Vansteenwegen P, Oudheusden DV (2007) Decreasing the passenger waiting time for an intercity rail network. Transp Res Part B 41:478–492
20. Ghoseiri K, Szidarovszky F, Asgharpour M (2004) A multi-objective train scheduling model and solution. Transp Res Part B 38:927–952
21. Hooke R, Jeeves T (1961) Direct search solution of numerical and statistical problems. J Assoc Comput Mach 8:212–229
22. Martí R (2010) Advaced multi-start methods. International series in operations research and management science: handbook of metaheuristics. Springer, New York, pp 265–281
23. The Mathworks Inc. (2004) Genetic algorithm and direct search toolbox for use with MATLAB: user's guide. The Mathworks Inc., Natick, MA, USA
24. Boggs P, Tolle J (1995) Sequential quadratic programming. Acta Numerica 4:1–51
25. Gill P, Murray W, Saunders M (2002) SNOPT: an SQP algorithm for large-scale constrained optimization. Soc Ind Appl Math J Optim 12:979–1006
26. The Mathworks Inc. (1999) Optimization toolbox for use with Matlab: user's guide. The Mathworks Inc., Natick, MA, USA
27. Williams H (1999) Model building in mathematical programming, 4th edn. Wiley, Chichester
28. Holmström K, Göran AO, Edvall M (2007) User's guide for tomlab/minlp. http://tomopt.com/docs/TOMLAB_MINLP.pdf
29. Berthold T, Gamrath G, Gleixner A, Heinz S, Koch T, Shinano Y (2012) Solving mixed integer linear and nonlinear problems using the SCIP Optimization Suite. ZIB-Report 12–17. Zuse Institute Berlin. Berlin, Germany

30. Kvasnica M, Grieder P, Baotić M (2011) Automatic derivation of optimal piecewise affine approximations of nonlinear systems. In: Proceedings of the 18th IFAC world congress. Milano, Italy, pp 8675–8680
31. Williams H (1993) Model building in mathematical programming. Wiley, New York
32. Linderoth J, Ralphs T (2005) Noncommercial software for mixed-integer linear programming. In: Karlof J (ed) Integer programming: theory and practice., Operations research seriesCRC Press, Boca Raton, pp 253–303
33. Atamturk A, Savelsbergh M (2005) Integer-programming software systems. Annals of operations research 140:67–124
34. Byrd R, Hribar M, Nocedal J (1999) An interior point algorithm for large-scale nonlinear programming. SIAM J Optim 9:877–900
35. Grant M, Boyd S (2012) CVX: Matlab software for disciplined convex programming, version 2.0 beta. http://cvxr.com/cvx
36. Zhang B, Lu Y (2011) Research on dwelling time modeling of urban rail transit. Traffic Transp 27:48–52

Chapter 6
OD-Dependent Train Scheduling for an Urban Rail Transit Line

Abstract In this chapter, in order to capture more detailed information about passengers we consider train scheduling with origin–destination-dependent (OD-dependent) passenger demands for an urban rail transit line. A stop-skipping strategy is adopted to reduce the total passenger travel time and the energy consumption. The resulting train scheduling problem is a mixed integer nonlinear programming problem. A bi-level approach and a limited bi-level approach are proposed to solve this problem. These two approaches are compared through a case study inspired by real data from the Beijing Yizhuang subway line. The results discussed in this chapter are based on Wang et al. (IEEE Trans Intell Transp Syst 15:2658–2670, 2014) [1]; Wang et al. (Proceedings of the 93rd annual meeting of the transportation research board, Washington, DC, 2014) [2].

6.1 Introduction

In Sect. 5.5, the train scheduling model has been extended such that trains can skip small stations with low passenger alighting proportions while assuming that the effect of skipping is negligible. Because OD-independent passenger demands are used in Chap. 5, the effect on the passengers who have the skipped stations as their destination is not taken into account in Sect. 5.5. Therefore, in this chapter, a train scheduling model based on OD-dependent passenger demands is presented. Since OD-dependent passenger demands vary significantly along the urban rail transit line and the time of the day, e.g., some stations (e.g., those in the central business district) may have a relatively large number of passengers boarding and alighting and others may have few passengers, fixed all-stop train schedules cannot efficiently satisfy such OD-dependent demand patterns. Hence, a dynamic stop-skipping strategy (i.e., the stop-skipping stations for each train are not fixed before the optimization process) based on OD-dependent passenger demands is important for both passengers and rail operators [1, 2].

This chapter is structured as follows. Section 6.2 formulates the operation of trains at the terminal station, at stations, and in between stations, and the passenger demand characteristics. Section 6.3 describes the multi-objective cost function of the train

Y. Wang et al., *Optimal Trajectory Planning and Train Scheduling for Urban Rail Transit Systems*, Advances in Industrial Control,
DOI 10.1007/978-3-319-30889-0_6

scheduling problem with stop-skipping. Section 6.4 proposes a bi-level optimization approach and a limited bi-level approach for the train scheduling problem with stop-skipping. Section 6.5 illustrates the performance of the proposed solution approaches with a case study. Finally, Sect. 6.6 concludes the chapter.

6.2 Model Formulation

This chapter considers an urban rail transit line as presented in Fig. 6.1, where the terminal station and stations in the line are numbered increasingly. Let J denote the total number of stations (terminal station not included). The index of the terminal station is set as 0.

The track section between station j and station $j+1$ is denoted as segment j. The scheduling time period for the train scheduling problem is denoted as $[t_0, t_{\text{end}}]$. To distinguish the different running cycles of the physical trains, so-called train services are introduced, where the train service number in a unique way identifies a train and its current cycle. When a physical train arrives at the terminal station, its service number will be augmented when this train departs. More specifically, the transit line has I physical trains in total, which are numbered as train $1, 2, \ldots, I$. However, the service number of trains increases with the cycle of the operation of trains. During the scheduling period, the service number of trains is $1, 2, \ldots, I, I+1, I+2, \ldots, 2I, \ldots, N_{\text{cyc}}I$, where N_{cyc} is number of the cycles for the operation of trains in a given time period $[t_0, t_{\text{end}}]$. The service number of a train increases with I when the train departs from the terminal station. Hence, train service i corresponds to physical train $[(i-1) \bmod I] + 1$. For the sake of simplicity, we use "train i" as a shorthand for "train service i" from now on.

We make the following assumptions[1] for the terminal station and the stations:

A.1 Multiple trains can be present at terminal station 0, which has a maximum capacity C_0^{ter}. In addition, the trains in terminal station 0 will depart from the terminal station in a first-in-first-out manner.

A.2 Station j for $j \in \{1, 2, \ldots J\}$ can only accommodate one train at a time and no overtaking can occur at any point of the line.

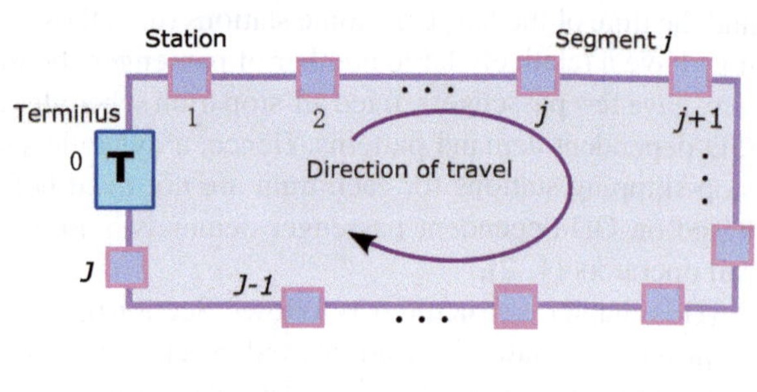

Fig. 6.1 Illustration of the urban rail transit line

[1] Assumption A.2, A.5, and A.7 are also considered in Chap. 5.

A.3 Trains can skip some stations in the urban transit line, where we define the skipping set $S_{\text{skip}} = \{(i, j) | \text{ train } i \text{ may potentially}^2 \text{ skip station } j\}$.
A.4 In view of preannouncement for passengers about the stop-skipping schedule, there is detailed stop-skipping information displayed in the station and/or the urban rail transit operator provides this information to passengers through mobile devices.
A.5 The operation of trains only consists of three phases: the acceleration phase, the speed holding phase, and the deceleration phase. Moreover, the acceleration and the deceleration are taken to be fixed constants.
A.6 Each passenger only takes one train to arrive at his/her destination, i.e., the transfer between different trains along the line is not allowed.
A.7 The number of passengers waiting at a station and the number of passengers on board immediately after a train's departure are approximated by real numbers.

Assumption A.1 can be motivated as follows: since we assume that there are multiple track sections in the terminal station, multiple trains can be present at the terminal station. Furthermore, a first-in-first-out operation for trains in the terminal station is not difficult to realize in practice since it depends on the dispatching of trains in the terminal station and it is a matter of renumbering of trains. Assumption A.2 generally holds for most urban transit systems, which are usually operated in this way [3, 4]. Even though the stop-skipping strategy is not yet widely used in urban rail transit networks throughout the world, there are several lines which apply it as mentioned before, e.g., the SEPTA line in Philadelphia. As a result, Assumption A.3 is possible in practice. With the recent development in information and communication technologies, it is possible to provide personalized transportation information to passengers via their mobile devices (like smart phone, PDAs, etc.). In addition, many stations in urban rail transit systems have screens to display travel information to passengers at platforms. Hence, Assumption A.4 is reasonable. In order to simplify the operation model for the trains, the detailed dynamics are not included in the model formulation, but only the three phases mentioned in Assumption A.5 are taken into account. However, once the running times between consecutive stations are fixed, a more accurate speed profile for the operation of trains can be calculated as a lower level control problem (see Chap. 3 for more information). Since we only consider a single line and since in Assumption A.2 we suppose that no overtaking can happen at any point of the line, the transfer between different trains for passengers is useless. In addition, if the headway between trains is small enough, then passengers will not choose to take the first arriving train and transfer to another train in the opposite direction to arrive at their destination since this costs extra transfer time and waiting time. Therefore, it is reasonable to assume that passengers will wait at their origin for the right train to get to their destination as is stated in Assumption A.6. For Assumption A.7, the number of passengers is usually large, so the error made by this assumption is small. Furthermore, this assumption simplifies the optimization problem later on.

---

[2] A binary variable will be introduced to indicate whether train i stops at station j or skips station j (see Sect. 6.2.1 for more details). The values of these binary variables will be optimized to minimize the objective function of the train scheduling problem.

Remark 6.1 When realizing the stop-skipping strategy in practice, rail transit operators should be aware that some passengers may not like it since they want to keep their journey as simple as possible, e.g., when a train arrives at the platform they want to broad the train. In addition, passenger could get on wrong trains which will not stop at their destinations, then they would feel frustrated. Furthermore, the passenger satisfaction to urban rail transit systems would go down because passengers waiting at the skipping stations will watch a train passing away without stop. Therefore, a sophisticated passenger information system is essential for the stop-skipping strategy.

6.2.1 Arrivals and Departures with Stop-Skipping

As has been indicated in Sect. 5.2.1, in order to obtain a balanced trade-off between the accuracy and the computation speed, a macroscopic model is used for the train scheduling. Section 5.2.1 has derived the arrival and departure equations for the operation of trains at stations. The operation of trains in the terminal station is also included. We will first formulate the operation of trains at the terminal station and then at the stations.

6.2.1.1 Operation of Trains in the Terminal Station

A train can depart from the terminal station only after it has arrived. Furthermore, the train number is increased with I when it departs from the terminal station. Hence, we have

$$d_{i,0} \geq a_{i-I,0} + \tau_{0,\min}, \tag{6.1}$$

where $d_{i,0}$ is the departure time of train i at the terminal station, a_{i-I} is the arrival time of train $i - I$, and $\tau_{0,\min}$ is the minimum dwell time for the trains at the terminal station. The minimum dwell time could be equal to the minimum turnaround time or the minimum running time at a terminal station. In addition, there is no upper bound for the dwell time of trains at the terminal station. Since we assume that there are multiple tracks in terminal station to accommodate trains, the running distance for trains between stations and the terminal station varies and depends on the route setting in the terminal station. However, the layout of the terminal station and the scheduling of trains in the terminal station are out of the scope of this chapter. We assume an average distance s_0 for trains running between the terminal station and station 1 and an average distance s_J for the trains running between station J and the terminal station. The arrival time of train i at the terminal station can then be written as

$$a_{i,0} = d_{i,J} + r_{i,J}, \tag{6.2}$$

where $r_{i,J}$ is the running time on segment J.

The headway constraints in the terminal station can be formulated as

$$d_{i,0} \geq d_{i-1,0} + h_{0,\mathrm{dep}}, \tag{6.3}$$

where $h_{0,\mathrm{dep}}$ is the minimum departure headway at the terminal station. In addition, the minimum arrival headway at the terminal station should also be taken into account, which can be formulated as

$$a_{i,0} \geq a_{i-1,0} + h_{0,\mathrm{arr}}, \tag{6.4}$$

where $h_{0,\mathrm{arr}}$ is the minimum arrival headway at the terminal station.

As mentioned in Assumption A.2, the capacity of the terminal station is C_0^{ter}. Therefore, at any time t the number of trains in the terminal station should be less than the capacity, which can be formulated as

$$\sum_{i \in S_{\mathrm{trains}}} \mathfrak{I}(a_{i,0} \leq t) - \sum_{i \in S_{\mathrm{trains}}} \mathfrak{I}(t \geq d_{i,0}) \leq C_0^{\mathrm{ter}}, \quad \text{for all } t$$

where S_{trains} is the set of trains considered in the scheduling problem and the indicator function $\mathfrak{I}(\cdot)$ is defined as

$$\mathfrak{I}(x) = \begin{cases} 1 \text{ if } x \text{ is true,} \\ 0 \text{ if } x \text{ is false.} \end{cases}$$

The number of trains in the terminal station only increases when a train arrives at the terminal station. Therefore, we should only check the capacity constraint when a train arrives at a terminal station; so the constraints can be reformulated as

$$\sum_{i \in S_{\mathrm{trains}}} \mathfrak{I}(a_{i,0} \leq a_{\ell,0}) - \sum_{i \in S_{\mathrm{trains}}} \mathfrak{I}(a_{\ell,0} \geq d_{i,0}) \leq C_0^{\mathrm{ter}}, \tag{6.5}$$

for each $\ell \in S_{\mathrm{trains}}$.

6.2.1.2 Operation of Trains at Stations

By Assumption A.3, trains can skip any station in the skipping set S_{skip}. If a train skips a station, then the train passes that station without stopping and the dwell time is then equal to zero. We have introduced a binary variable in Sect. 5.5 to indicate whether a train will stop at a station or not (see 5.28):

$$y_{i,j} = \begin{cases} 1 \text{ if train } i \text{ will stop at station } j, \\ 0 \text{ if train } i \text{ will skip station } j. \end{cases}$$

In Sect. 5.2.1, the departure time $d_{i,j}$ of train i at station j is equal to the sum of the arrival time $a_{i,j}$ and the dwell time $\tau_{i,j}$ of train i at station j as shown in (5.1). We reformulate (5.1) as the following two inequalities, i.e., the departure time $d_{i,j}$ should satisfy

$$d_{i,j} \geq a_{i,j} + y_{i,j}\tau_{i,j,\min}, \tag{6.6}$$

and

$$d_{i,j} \leq a_{i,j} + y_{i,j}\tau_{i,j,\max}, \tag{6.7}$$

where $a_{i,j}$ is the arrival time of train i at station j, the minimum dwell time $\tau_{i,j,\min}$ is influenced by the number of passengers boarding and alighting from the train (see Sect. 5.2.3 for more details), and the maximum dwell time $\tau_{i,j,\max}$ is introduced to ensure passenger satisfaction. The arrival time $a_{i,j+1}$ of train i at station $j+1$ can be calculated by (5.2) in 5.2.1. The running time $r_{i,j}$ should be calculated by (5.3.2) instead of (5.3). Furthermore, the minimum headway is the same as (5.5). Moreover, if train i skips station $j+1$, i.e., $y_{i,j+1} = 0$, then the operation of the train between station j and station $j+2$ only contains three phases and the holding speed should satisfy (5.31) in Sect. 5.5.

6.2.2 OD-Dependent Passenger Demand Characteristics

Section 5.2.2 has described OD-independent passenger demand characteristics. Passenger demands with origin and destination are more complex than that in Sect. 5.2.2. The relationship between the variables used for describing the OD-dependent passenger characteristics is illustrated in Fig. 6.2. The number of waiting passengers $w_{i,j}^{\text{wait}}$ for train i at station j is equal to the sum of the number of waiting passengers $w_{i,j}^{\text{wait}}$ with destination m for all $m \in \{j+1, j+2, \ldots, m, \ldots, J\}$, i.e.,

$$w_{i,j}^{\text{wait}} = \sum_{m=j+1}^{J} w_{i,j,m}^{\text{wait}}. \tag{6.8}$$

The number of waiting passengers $w_{i,j}^{\text{wait}}$ with destination m can be calculated by

$$w_{i,j,m}^{\text{wait}} = w_{i-1,j,m} + \lambda_{j,m}(d_{i,j} - d_{i-1,j}), \tag{6.9}$$

where $w_{i-1,j,m}$ is the number of passengers with destination station m remaining at station j immediately after the departure of train $i-1$, $\lambda_{j,m}(d_{i,j} - d_{i-1,j})$ is the number of newly arrived passengers in between the departures of train i and train $i-1$, and $\lambda_{j,m}$ is the passenger arrival rate at station j for passengers with destination m. Note that the passenger arrival rate at the final station J, is assumed to be zero since we only consider one direction of the line.

The number of passengers alighting from train i at station j is denoted as $n_{i,j}^{\text{alight}}$ and it can be computed using

$$n_{i,j}^{\text{alight}} = \sum_{\ell=1}^{j-1} n_{i,\ell,j}^{\text{board}}, \tag{6.10}$$

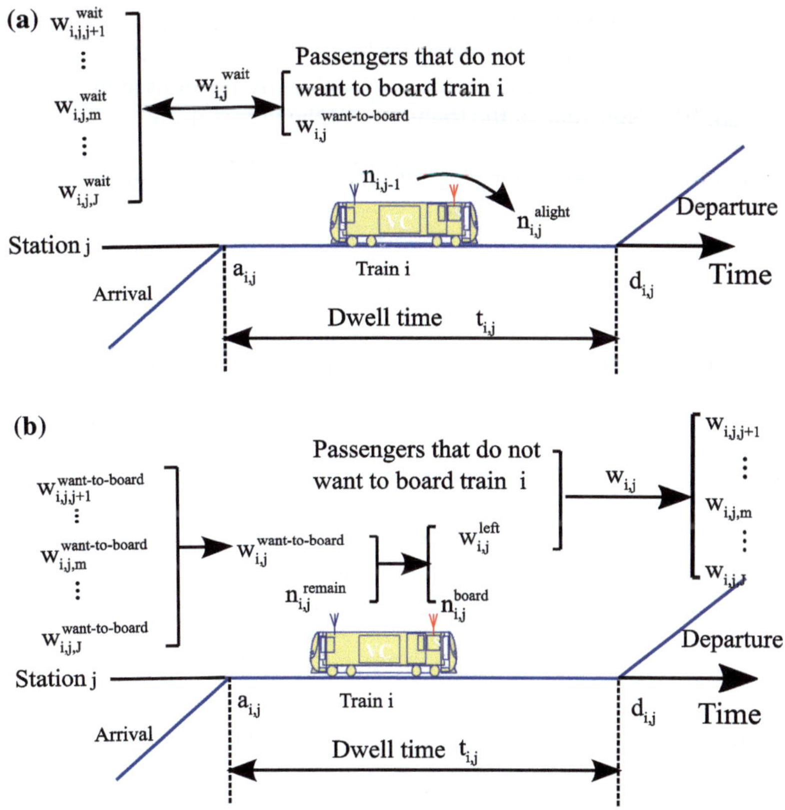

Fig. 6.2 The variables for the passenger characteristics. **a** The alighting process of passengers at station j. **b** The boarding process of passengers at station j

where $n^{\text{board}}_{i,\ell,j}$ is the number of passengers that have station j as their destination and have boarded train i at station ℓ. The number of passengers $n^{\text{board}}_{i,\ell,j}$ can be calculated by

$$n^{\text{board}}_{i,\ell,j} = w^{\text{wait}}_{i,\ell,j} - w_{i,\ell,j}, \tag{6.11}$$

where $w_{i,\ell,j}$ is the number of passengers who are left by train i at station ℓ and have station j as their destination. The calculation for $w_{i,\ell,j}$ will be given in (6.17). Note that no passenger will get off the train if train i skips station j because the passengers at upstream stations were informed that train i would not stop at station j; so in that case those passengers with station j as destination would not get on train i. The number of passengers who want to board train i at station j and have station m as their destination is denoted as $w^{\text{want-to-board}}_{i,j,m}$. The number of passengers $w^{\text{want-to-board}}_{i,j,m}$ relies on whether train i stops at station j and whether train i stops at station m for $m \in \{j+1, j+2, \ldots, J\}$, i.e.,

$$w^{\text{want-to-board}}_{i,j,m} = y_{i,j} y_{i,m} w^{\text{wait}}_{i,j,m}. \tag{6.12}$$

So if train i skips station j, i.e., $y_{i,j} = 0$, then no passengers want to board train i, i.e., $w_{i,j}^{\text{want-to-board}} = 0$. If train i stops at station j, i.e., $y_{i,j} = 1$, then the number of passengers who want to board is decided by whether train i stops at their destination m, i.e., $y_{i,m} w_{i,j,m}^{\text{wait}}$. Note that all the trains stop at the terminal station, so $y_{i,0}$ is equal to 1 for $i \in \{1, 2, \ldots, I\}$.

The number of passengers $w_{i,j}^{\text{want-to-board}}$ who want to board train i at station j is

$$w_{i,j}^{\text{want-to-board}} = \sum_{m=j+1}^{J} w_{i,j,m}^{\text{want-to-board}}. \tag{6.13}$$

The number of passengers on train i immediately after its departure at station j is defined as $n_{i,j}$, which can be computed as

$$n_{i,j} = n_{i,j-1} - n_{i,j}^{\text{alight}} + n_{i,j}^{\text{board}}, \tag{6.14}$$

where the number of boarding passengers $n_{i,j}^{\text{board}}$ equals the minimum of the number of passengers $w_{i,j}^{\text{want-to-board}}$ that want to board train i and the remaining capacity $n_{i,j}^{\text{remain}}$ of the train i at station j immediately after the alighting process:

$$n_{i,j}^{\text{board}} = \min(n_{i,j}^{\text{remain}}, w_{i,j}^{\text{want-to-board}}). \tag{6.15}$$

Moreover, the number of passengers $n_{i,j}^{\text{board}}$ boarding train i at station j is equal to

$$n_{i,j}^{\text{board}} = \sum_{m=j+1}^{J} n_{i,j,m}^{\text{board}}.$$

Furthermore, the remaining capacity of train i at station j immediately after the alighting process is

$$n_{i,j}^{\text{remain}} = C_{\max} - n_{i,j-1} + n_{i,j}^{\text{alight}}. \tag{6.16}$$

The number of passengers $w_{i,j}^{\text{left}}$ left by train i is decided by whether train i will stop at station j or not. We have the following two cases:

- Train i skips station j, i.e., $y_{i,j} = 0$
  If train i will skip station j, then the number of boarding passengers $n_{i,j}^{\text{board}}$ equals zero. All the passengers waiting at station j will then be left by train i.
- Train i will stop at station j, i.e., $y_{i,j} = 1$
  If $w_{i,j}^{\text{want-to-board}} \leq n_{i,j}^{\text{remain}}$, then all the passengers that want to board can get on train i. However, if $w_{i,j}^{\text{want-to-board}} > n_{i,j}^{\text{remain}}$ there will be passengers left by train i at station j. The number of passengers $w_{i,j}^{\text{left}}$ who want to board but cannot get on train i at station j immediately after the departure of train i can be calculated by

$$w_{i,j}^{\text{left}} = w_{i,j}^{\text{want-to-board}} - \min(n_{i,j}^{\text{remain}}, w_{i,j}^{\text{want-to-board}}) \quad \text{if } y_{i,j} = 1.$$

In this case, if train i stops station m for $m \in \{j+1, j+2, \ldots, J-1\}$, i.e., $y_{i,m} = 1$, we assume that the number of passengers that have station m as destination and are left by train i is proportional to the number of passengers who want to board. The number of passengers who have destination m and are left by train i can be formulated as

$$w_{i,j,m} = w_{i,j}^{\text{left}} \frac{w_{i,j,m}^{\text{want-to-board}}}{w_{i,j}^{\text{want-to-board}}} \quad \text{if } y_{i,j} = 1 \text{ and } y_{i,m} = 1.$$

However, if train i skips station m for $m \in \{j+1, j+2, \ldots, J-1\}$, i.e., $y_{i,m} = 0$, then the number of passengers that have station m as destination will not board. So we have

$$w_{i,j,m} = w_{i,j,m}^{\text{wait}}, \quad \text{if } y_{i,j} = 1 \text{ and } y_{i,m} = 0.$$

Therefore, the number of passengers who are left by train i and with destination m can be calculated as

$$w_{i,j,m} = y_{i,j}\left(y_{i,m} w_{i,j}^{\text{left}} \frac{w_{i,j,m}^{\text{want-to-board}}}{w_{i,j}^{\text{want-to-board}}} + (1 - y_{i,m}) w_{i,j,m}^{\text{wait}} \right) + (1 - y_{i,j}) w_{i,j,m}^{\text{wait}}. \quad (6.17)$$

Furthermore, the total number of waiting passengers at station j immediately after the departure of train i is

$$w_{i,j} = \sum_{m=j+1}^{J} w_{i,j,m}. \quad (6.18)$$

Remark 6.2 The minimum dwell time for the passengers to get on and get off trains can be calculated using the linear model (5.11) and the nonlinear model (5.12) presented in Sect. 5.2.3. See there for more information.

6.3 Mathematical Formulation of the Scheduling Problem

The train scheduling problem with stop-skipping is a multi-objective optimization problem. In this chapter, the objective is similar as (5.21) defined in Sect. 5.3, where the energy consumption of trains and the travel time of passengers are taken into account.

For the train scheduling problem with stop-skipping, the energy consumption of trains can be calculated by (5.17), (5.18), and (5.35) in Sect. 5.3. The total travel time can be calculated by (5.16), which is repeated as follows:

$$t_{\text{total}} = \sum_{i=1}^{I} \sum_{j=1}^{J-1} (\gamma_{\text{wait}} t_{\text{wait},i,j} + t_{\text{in-vehicle},i,j}).$$

However, since we consider OD-dependent passenger demands in this chapter, instead of (5.14) and (5.15), the computation for the passenger waiting time and passenger in-vehicle time should be calculated as follows:

$$t_{\text{wait},i,j} = w_{i-1,j}(d_{i,j} - d_{i-1,j}) + \frac{1}{2} \sum_{m=j+1}^{J} \lambda_{j,m}(d_{i,j} - d_{i-1,j})^2, \tag{6.19}$$

and

$$t_{\text{in-vehicle},i,j} = n_{i,j}r_{i,j} + (n_{i,j} - n_{i,j}^{\text{alight}})\tau_{i,j+1}. \tag{6.20}$$

In order to spread trains over the entire scheduling time period, a penalty term is introduced to describe the waiting time of the passengers left by the last train $N_{\text{cyc}}I$ during the scheduling period:

$$f_{\text{penalty},1} = \sum_{j=1}^{J} \left(w_{N_{\text{cyc}}I,j}(t_{\text{end}} - d_{N_{\text{cyc}}I,j}) + \frac{1}{2}\zeta \sum_{m=j+1}^{J} \lambda_{j,m}(t_{\text{end}} - d_{N_{\text{cyc}}I,j})^2 \right). \tag{6.21}$$

If $\zeta = 1$, the waiting time for the newly arrived passengers between the departure time $d_{N_{\text{cyc}}I,j}$ and the end time t_{end} is also taken into account. However, if $\zeta = 0$, the waiting time of the newly arrived passengers after the last train is not considered (e.g., the trains coming later will pick up these passengers). But in the latter case, another penalty term should be added for the arrival time of the last train $N_{\text{cyc}}I$ at the terminal station in order to avoid all the trains operating close to each other at the start of the period $[t_0, t_{\text{end}}]$:

$$f_{\text{penalty},2} = |a_{N_{\text{cyc}}I,0} - t_{\text{end}}|. \tag{6.22}$$

The objective function of the train scheduling problem can be written as

$$f_{\text{opt}} = \gamma_1 \frac{E_{\text{total}}}{E_{\text{total,nom}}} + \gamma_2 \frac{t_{\text{total}}}{t_{\text{total,nom}}} + \mu_{\text{penalty},1} \frac{f_{\text{penalty},1}}{f_{\text{penalty},1,\text{nom}}} + \mu_{\text{penalty},2} \frac{f_{\text{penalty},2}}{f_{\text{penalty},2,\text{nom}}}, \tag{6.23}$$

where γ_1, γ_2, $\mu_{\text{penalty},1}$, and $\mu_{\text{penalty},2}$ are nonnegative weights, and the normalization factors $E_{\text{total,nom}}$, $t_{\text{total,nom}}$, $f_{\text{penalty},1,\text{nom}}$, and $f_{\text{penalty},2,\text{nom}}$ are "nominal" values of the total energy consumption and the total travel time of passengers, respectively. These nominal values can, e.g., be determined by running trains according to a feasible initial schedule.

Remark 6.3 A rolling horizon approach can be adopted to solve the train scheduling problem. See Sect. 7.4 for more details on this.

6.4 Solution Approaches

The resulting train scheduling problem is a mixed integer nonlinear programming (MINLP) problem with objective function (6.23) and constraints (6.1)–(6.18). The objective function (6.23) is nonlinear and nonconvex. In addition, we have the non-smooth min function in constraints (6.14), (6.15), and (6.17). This MINLP problem can be solved using, e.g., the branch-and-bound method, but the computation time is quite long even for small-sized problems in practice. We could also approximate the nonlinear terms in the MINLP problem using piecewise affine functions and transform the MINLP problem into a mixed integer linear programming (MILP) problem. However, we need to approximate many nonlinear terms of the scheduling problem; so the performance of the MILP approach may be not good anymore because of the approximation errors.

In this section, we propose a bi-level optimization approach to solve the optimization problem under consideration. Moreover, we introduce a limited bi-level solution approach for the MINLP problem to reduce the computation time, where the search space of the problem is limited and a threshold method is presented to obtain a good initial solution for the MINLP problem.

6.4.1 Bi-Level Optimization Approach

The free variables in the real-time scheduling problem are the departure times $d_{i,j}$, the holding speeds $v_{i,j}$, and the binary variables $y_{i,j}$ for all trains and stations. The other variables like the number of passengers waiting at stations $w_{i,j}$ and the number of passengers on board the trains $n_{i,j}$ can be eliminated using the model equations (6.8)–(6.18). The proposed bi-level optimization method consists of two levels of optimization:

- The high-level optimization determines the binary variables $y_{i,j}$ (only if $(i, j) \in S_{\text{skip}}$ $y_{i,j}$ could be equal to 0), where a brute force approach can be used to explore all the combinations for the binary variables in case the size of the problem is small. Alternatively, integer programming approaches, such as branch-and-bound and heuristics, can be applied in the high-level optimization if the size of the problem is big.
- For each combination of binary variables, the low-level optimization solves a nonlinear nonconvex problem using, e.g., multi-start sequential quadratic programming (SQP) algorithm [5] or a pattern search method [6].

The procedure of the bi-level optimization method is given in Algorithm 2, where—for illustration purpose—the high-level optimization problem is solved using a genetic algorithm and a multi-start optimization algorithm is used in the lower level. The feasibility of the low-level optimization problem depends on the value of the binary variables. If the low-level optimization problem is infeasible, we introduce

Algorithm 2 Bi-level optimization approach for the train scheduling problem with stop-skipping

Input : maximum number of generations G, population size s_{p}, initial population P_0 of the binary variables, number of initial points $k_{\max}$ used in the low-level optimization, a large value F for the fitness function to indicate infeasibility, feasible binary solution δ_{f}, positive weight λ_{f};
for $g = 0, 1, \ldots, G-1$ **do**
  for $\ell = 1, 2, \ldots, s_{\mathrm{p}}$ **do**
    binary variables $\delta \leftarrow \ell$-th parent $P_g(\ell)$ from the g-th generation;
    for $k = 1, 2, \ldots, k_{\max}$ **do**
      generate an initial random feasible solution $d_{i,j}(\ell, 0)$, $r_{i,j}(\ell, 0)$, $\tau_{i,j}(\ell, 0)$, $w_{i,j}(\ell, 0)$, and $n_{i,j}(\ell, 0)$ for $i = 1, \ldots, I$ and $j = 1, \ldots, J$;
      if low-level optimization problem turns out to be feasible based on current values of δ and initial solutions **then**
        $d_{i,j}(\ell, k)$, $r_{i,j}(\ell, k)$, $\tau_{i,j}(\ell, k)$, $w_{i,j}(\ell, k)$, $n_{i,j}(\ell, k)$, and $f_{\mathrm{opt}}(\ell, k) \leftarrow$ solution of the low-level optimization problem;
      else
        $f_{\mathrm{opt}}(\ell, k) \leftarrow F + \lambda_{\mathrm{f}} \|\delta - \delta_{\mathrm{f}}\|_2$;
      end if
    end for
    value of fitness function $f^*_{\mathrm{opt}}(\ell)$ for the ℓ-th parent $\leftarrow \min_{k=1,\ldots,k_{\max}} f_{\mathrm{opt}}(\ell, k)$;
  end for
  select new parents from the current population based on the fitness function f_{opt};
  generate a new generation population of binary variables through crossover and mutation;
end for
Output : choose the best offspring solution at generation G and calculate $d_{i,j}$, $r_{i,j}$, $\tau_{i,j}$, $w_{i,j}$, and $n_{i,j}$.

a large penalty value F for the fitness function as shown in Algorithm 2. Furthermore, in order to steer infeasible binary variables toward feasible ones, we also add the norm of $\delta - \delta_{\mathrm{f}}$ in the fitness function where δ_{f} is a feasible value of δ and can be decided by using the known information about the urban rail transit line, e.g., passenger flows and line conditions.

6.4.2 *Limited Bi-Level Optimization Approach*

In the limited bi-level optimization problem, we propose a threshold method to obtain good initial solutions for the MINLP problem. In addition, we limit the search space of train scheduling problem in the neighborhood of these initial solutions.

6.4.2.1 Threshold Method for Obtaining Good Initial Solutions

First, a threshold function is introduced to determine the values of the stopping variables in order to get a good initial solution for the train scheduling problem. The threshold function is defined as follows:

$$y_{i,j} = \mathfrak{I}\Big(\big(w_{i,j}^{\text{want-to-board}} \geq \theta_{i,j}^{\text{th,in}}\big) \wedge \big(n_{i,j}^{\text{alight}} \geq \theta_{i,j}^{\text{th,out}}\big)\Big), \tag{6.24}$$

where $\mathfrak{I}(\cdot)$ is the indicator function (see Sect. 6.2.1 for more information), $\theta_{i,j}^{\text{th,in}}$ and $\theta_{i,j}^{\text{th,out}}$ are the thresholds, which are free variables and determined by the optimization procedure. In this case, the value of the stopping variables depends on the passenger flows in the urban rail transit line. By introducing the threshold function, we can reformulate the MINLP problem as a real-valued nonlinear programming problem, which can be solved by, e.g., a SQP method. The resulting solution then yields an initial solution of the stopping variables, the departure times, and the holding speeds for the original, full MINLP problem.

6.4.2.2 Limiting the Search Space

Based on an initial solution obtained by the threshold method, we can limit the search space of the train scheduling problem within a neighborhood of the initial solution to reduce the computation time. For the high-level optimization, the search space of the integer variables $y = \left[y_{1,1} y_{1,2}, \ldots, y_{2,1}, \ldots, y_{N_{\text{cyc}I},J}\right]^T$ can be limited by a 1-norm constraint:

$$\|y - y_{\text{init}}\|_1 \leq \chi_0, \tag{6.25}$$

which means that only a limited number, i.e., χ_0, of binary variables can change their values in the bi-level optimization approach. If χ_0 is small, the search space can be reduced dramatically. If χ_0 is chosen as 1 or 2 and the number of trains and stations is not too large, a brute force method could be applied for the high-level optimization. Otherwise, a genetic algorithm could be employed.

Moreover, we can also limit the search space of the departure times and holding speeds as follows [7, 8]:

$$\|d - d_{\text{init}}\|_2 \leq \chi_2 + \chi_3\, \beta, \tag{6.26}$$

and

$$\|v - v_{\text{init}}\|_2 \leq \chi_3 + \chi_4\, \beta, \tag{6.27}$$

and the objective function (6.23) is revised as

$$f'_{\text{opt}} = f_{\text{opt}} + \gamma_4\, \beta, \tag{6.28}$$

where β is a slack variable which introduces an element of slackness into the problem to make sure the resulting optimization problem is always feasible. The relative degree of under- or overachievement of the goals is controlled by the weights χ_3, χ_4 and γ.

6.5 Case Study

6.5.1 Setup

To demonstrate the effectiveness of the proposed model formulation and the performance of the proposed limited bi-level optimization approach, we consider a cyclic line with 1 terminal station and 12 stations following the structure given in Fig. 6.1. There are 6 physical trains in the cyclic line and the number of train services considered in the train scheduling problem is 10. The train characteristics and the line data are inspired by the data of Beijing Yizhuang subway line, and are given in Tables 6.1 and 6.2. The mass of the train, the mass of one passenger, and the coefficients for the minimum dwell time in (5.12) are given in Table 6.1. In Table 6.2, station 0 represents the terminal station. The minimum running time in Table 6.2 is calculated by taking a fixed acceleration of 0.8 m/s^2 and a fixed deceleration of -0.8 m/s^2; furthermore, when calculating the minimum running time the trains are supposed to run at the maximum speed of 22.2 m/s during the holding phase. The maximum running time is assumed to be $r_{i,j,\max} = \zeta r_{i,j,\min}$, where ζ is larger than 1. We have chosen ζ as 1.2 to guarantee that the passengers do not complain that the train is too slow. Note that we use the nonlinear model (5.12) for the calculation of the minimum dwell time, where the minimum dwell time $\tilde{\tau}_{\min}$ predefined by the rail operator in (5.13) is chosen as 30 s. The OD-dependent passenger arrival rates at stations are shown in Table 6.3.

The initial states at time t_0 (chosen as 1300 s for this case study) of the trains are as follows: train 1 and 2 are running to station 8 and 5, respectively. Because it is assumed that the schedule of a train can only be changed at stations, the arrival times of these two trains at station 3 and station 2 are fixed and they are 1400 and

Table 6.1 Parameters of the trains and the passengers [1]

Property	Symbol	Value
Train mass (kg)	$m_{e,i}$	1.99×10^5
Mass of one passenger (kg)	m_p	60
Capacity of trains (passengers)	$C_{i,\max}$	1.468×10^3
Minimum dwell time (s)	$\tilde{\tau}_{\min}$	30
Maximum dwell time (s)	$\tau_{\max}$	150
Coefficients of the minimal dwell time (s)	$\alpha_{1,d}$	4.002
(s/passengers)	$\alpha_{2,d}$	0.047
(s/passengers)	$\alpha_{3,d}$	0.051
(s/passengers^{-4})	$\alpha_{4,d}$	1.0×10^{-6}
Coefficients of resistance (m/s^2)	k_{1i}	0.012
(s^{-1})	k_{2i}	5.049×10^{-4}
(m^{-1})	k_{3i}	2.053×10^{-5}

Table 6.2 Information of the cyclic line for the case study

Station number	0	1	2	3	4	5	6	7	8	9	10	11	12
Distance to next station (m)	1050	1832	1786	2086	2265	1030	1354	1280	1544	992	1975	2369	1349
Minimal running time (s)	75.030	110.223	108.153	121.654	129.710	74.130	88.711	85.381	97.262	72.420	116.659	134.391	88.486

Table 6.3 Origin–destination-dependent passenger arrival rates (passengers/s) at stations [1]

Station	1	2	3	4	5	6	7	8	9	10	11	12
1	0	0.06	0.30	0.35	0.03	0.18	0.36	0.06	0.34	0.27	0.03	0.12
2	0	0	0.05	0.06	0.02	0.01	0.04	0.02	0.02	0.01	0.02	0.03
3	0	0	0	0.2	0.03	0.27	0.18	0.02	0.25	0.17	0.03	0.36
4	0	0	0	0	0.02	0.25	0.25	0.04	0.22	0.32	0.02	0.34
5	0	0	0	0	0	0.05	0.02	0.01	0.01	0.04	0.01	0.03
6	0	0	0	0	0	0	0.21	0.08	0.24	0.27	0.05	0.39
7	0	0	0	0	0	0	0	0.02	0.35	0.23	0.03	0.34
8	0	0	0	0	0	0	0	0	0.03	0.04	0.02	0.05
9	0	0	0	0	0	0	0	0	0	0.27	0.03	0.39
10	0	0	0	0	0	0	0	0	0	0	0.03	0.35
11	0	0	0	0	0	0	0	0	0	0	0	0.06
12	0	0	0	0	0	0	0	0	0	0	0	0

1340 s, respectively. Train 3 has arrived at 1270 s and it is stopping at station 3 at time t_0. The numbers of passengers on train 1, 2, and 3 at time t_0 and their destination are given in Table 6.4. In addition, there are 3 trains at the terminal station, so the corresponding previous train services finished before t_0. A communication-based train control system (a moving block signaling system) is implemented in Beijing Yizhuang subway line, where the minimum headway between two successive trains is 90 s. Additionally, a maximum departure-departure headway is considered to ensure the passenger satisfaction, which is chosen as 400 s. Furthermore, the numbers of passengers waiting at the various stations at t_0 and the destination of these passengers are shown in Table 6.5. The nominal values for the total travel time, the energy consumption, and the waiting time for the passengers who did not travel in the scheduling period are calculated according to a schedule with constant headway; they are 2.278×10^7 s, 7.013×10^9 J, and 1.387×10^7 s, respectively.

6.5.2 Results and Discussion

The train scheduling problem is solved by the following three approaches:

- All-stop approach. Trains in the scheduling period stop at every station, i.e., there is no skipping at all. In this case, the train scheduling problem is a nonlinear programming problem, which is solved here using the SQP method implemented by the fmincon function of MATLAB® optimization toolbox.

Table 6.4 Number of passengers on train 1, 2, and 3 at time t_0 and their destination

Destination station	1	2	3	4	5	6	7	8	9	10	11	12	Total number of passengers
Train 1	0	0	0	0	0	0	0	0	130	396	264	133	923
Train 2	0	0	0	0	0	89	35	111	46	333	166	22	802
Train 3	0	0	0	106	100	20	146	218	31	103	145	21	886

Table 6.5 Number of passengers waiting at stations at t_0 and their destination

Destination station	1	2	3	4	5	6	7	8	9	10	11	12	Total number of passengers
Station 1	0	8	75	52	65	15	26	32	15	62	33	21	417
Station 2	0	0	19	19	19	14	15	13	15	13	11	14	152
Station 3	0	0	0	29	92	15	45	32	12	41	33	14	313
Station 4	0	0	0	0	57	22	52	43	21	11	24	35	265
Station 5	0	0	0	0	0	13	22	29	26	14	26	24	154
Station 6	0	0	0	0	0	0	25	23	26	5	30	14	123
Station 7	0	0	0	0	0	0	0	23	25	13	27	29	117
Station 8	0	0	0	0	0	0	0	0	25	9	19	18	71
Station 9	0	0	0	0	0	0	0	0	0	24	20	23	67
Station 10	0	0	0	0	0	0	0	0	0	0	21	25	46
Station 11	0	0	0	0	0	0	0	0	0	0	0	26	26
Station 12	0	0	0	0	0	0	0	0	0	0	0	0	0

- Bi-level approach with stop-skipping. The train scheduling problem with stop-skipping is a MINLP problem, which is solved using the bi-level approach. A genetic algorithm is adopted for the integer optimization of the high level, where the ga function of the global optimization toolbox of MATLAB is employed. The nonlinear optimization problem in the lower level is solved using the SQP algorithm of the fmincon function of the MATLAB optimization toolbox.
- Limited bi-level approach with stop-skipping. First, a good initial solution for the mixed integer nonlinear problem is obtained by solving the train scheduling problem with threshold function (6.24), which is a nonlinear programming problem and which is solved by the SQP algorithm of the fmincon function of the MATLAB optimization toolbox. Based on an initial train schedule, we assign the value of χ_0 in (6.25) as 1, 2, and 3 to vary the search space. For $\chi_0 = 1$, we apply a brute force approach for the high-level optimization since the search space of the binary variables is small then. For $\chi_0 = 2$ or 3, the genetic algorithm function ga of the global optimization toolbox of MATLAB is used to optimize the binary variables.

6.5.2.1 Results

The train schedules obtained by the all-stop approach, the bi-level approach, and the limited bi-level approach are shown in Figs. 6.3, 6.4, 6.5, 6.6, 6.7, and 6.8. Although these train schedules look similar to each other, there are still differences between them. More specifically, for the all-stop approach (Fig. 6.3) all trains stop at all stations, while several trains skip some stations in the train schedules generated by the bi-level approach and the limited bi-level approach (see Figs. 6.4, 6.5, 6.6, 6.7 and 6.8). In the train schedule obtained by the bi-level approach shown in Fig. 6.4, trains 4, 6, 7, 8, 9, and 10 skip some stations. In particular, train 4 skips stations 2, 5, 8, and 11; so the stopping variables of train 4 for these stations are equal to 0 as given in Table 6.6. Additionally, it can be observed that the travel time for trains that skip some stations is smaller than that of the all-stop approach, e.g., train 4 arrives earlier at the terminal station in Fig. 6.4 (stop-skipping approach) than in Fig. 6.3 (all-stop

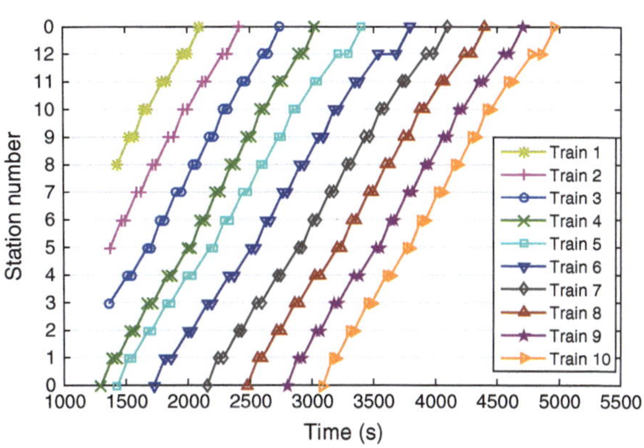

Fig. 6.3 Train schedule obtained by the all-stop approach

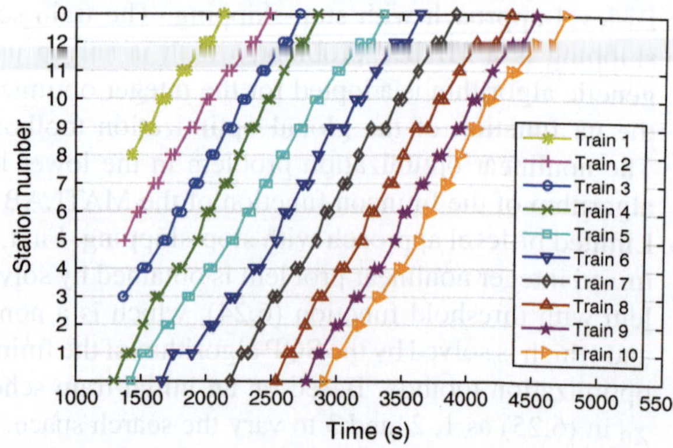

Fig. 6.4 Train schedule obtained by the bi-level approach

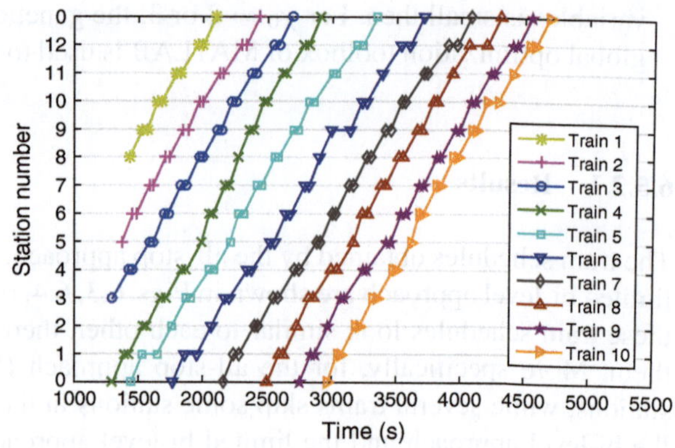

Fig. 6.5 Train schedule obtained by the threshold method

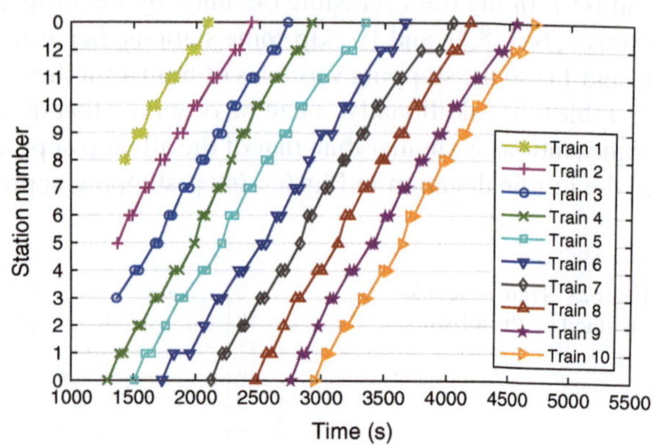

Fig. 6.6 Train schedule obtained by the limited bi-level approach with $\chi_0 = 1$

approach). Trains 1, 2, and 3 have already departed from the terminal station at time t_0 and they are thus assumed to stop at all stations. So the stopping variables for these three trains are equal to 1. Figure 6.5 illustrates an initial train schedule that is obtained by the threshold method. The values of the stopping variables are shown

Fig. 6.7 Train schedule obtained by the limited bi level approach with $\chi_0 = 2$

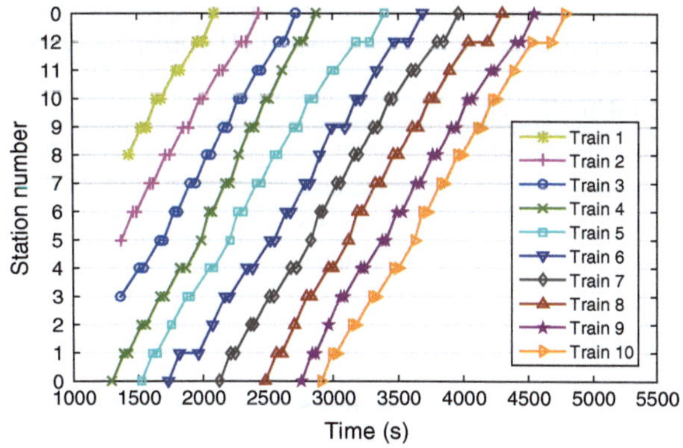

Fig. 6.8 Train schedule obtained by the limited bi-level approach with $\chi_0 = 3$

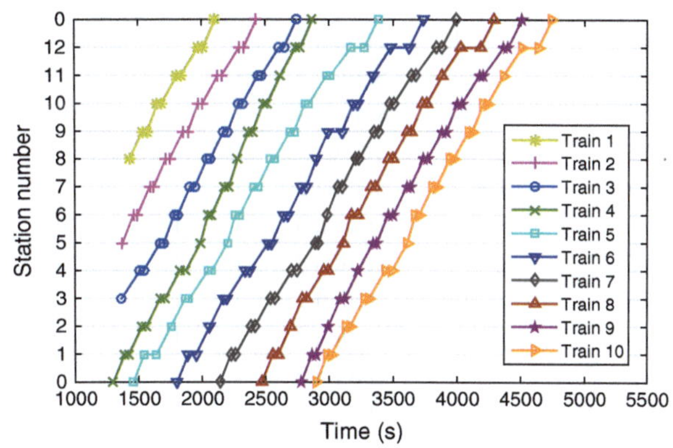

in Table 6.6, which are different from those obtained via the bi-level approach. For example, train 4 only skips stations 5 and 8 but does not skip station 2 and 11. Figure 6.6 shows the train schedule obtained by the limited bi-level approach with $\chi_0 = 1$, which means that one binary variable can change its value. From Fig. 6.6 and Table 6.6, it can be observed that train 7 skips station 5 when $\chi_0 = 1$, which is different from the initial schedule. Similarly, when $\chi_0 = 2$, the values of two binary variables are changed compared to the initial solution. As we can observe from Fig. 6.7 and Table 6.6, train 4 skips station 11 and train 7 skips station 5, while in the initial train schedule shown in Fig. 6.5, train 4 stops at station 11 and train 7 stops at station 5. Figure 6.8 presents the train schedule obtained by the limited bi-level approach with $\chi_0 = 3$, where train 4 skips station 11 and both train 7 and train 9 skips station 5 compared with the initial schedule.

Table 6.7 lists the values of the objective function, the computation time (on a 64-bit Linux operation system running on a 1.8 GHz Intel Core2 Duo CPU), the total passenger travel time, the energy consumption of the train schedules, etc., for each of the three approaches. Note that the comparison here is fair since all the approaches are implemented in M-files in MATLAB. The relative improvements of the bi-level

Table 6.6 Stopping variables of the all-stop approach, the bi-level approach, and the limited bi-level approach at station 2, 5, 8, and 11 [1] (The stopping variables for other stations are all equal to 1. Note that if the stopping variable equals 1, then the train stops at the station; otherwise, the train skips the station.)

Solution approach		Train	1	2	3	4	5	6	7	8	9	10
Bi-level		Station 2	1	1	1	0	1	1	0	0	1	1
		Station 5	1	1	1	0	1	0	0	0	0	0
		Station 8	1	1	1	0	1	0	0	0	1	0
		Station 11	1	1	1	0	1	0	0	0	0	1
Limited bi-level	Threshold	Station 2	1	1	1	1	0	0	1	0	0	1
		Station 5	1	1	1	0	0	1	1	0	1	0
		Station 8	1	1	1	0	1	0	1	1	1	1
		Station 11	1	1	1	1	1	0	1	0	1	0
	$\chi_0 = 1$	Station 2	1	1	1	1	0	0	1	0	0	1
		Station 5	1	1	1	0	0	1	0	0	1	0
		Station 8	1	1	1	0	1	0	1	1	1	1
		Station 11	1	1	1	1	1	0	1	0	1	0
	$\chi_0 = 2$	Station 2	1	1	1	1	0	0	1	0	0	1
		Station 5	1	1	1	0	0	1	0	0	1	0
		Station 8	1	1	1	0	1	0	1	1	1	1
		Station 11	1	1	1	0	1	0	1	0	1	0
	$\chi_0 = 3$	Station 2	1	1	1	1	0	0	1	0	0	1
		Station 5	1	1	1	0	0	1	0	0	0	0
		Station 8	1	1	1	0	1	0	1	1	1	1
		Station 11	1	1	1	0	1	0	1	0	1	0

approach and the limited bi-level approach with respect to the all-stop approach are given as in Table 6.8, and they are calculated as

$$x_{\text{relative-difference}} = 1 - \frac{x_{\text{stop-skipping}}}{x_{\text{all-stop}}},$$

where x is the value of the objective function, the computation time, etc., in the table and the stop-skipping involves the value obtained for the bi-level approach or the limited bi-level approach. Note that in Table 6.7, a solution approach has a better performance when the number of passengers finished their trips is larger since more passengers have finished their trip during the scheduling period. For the other criteria in Table 6.7, the performance of the approach is better if these criteria have a smaller value. In Table 6.8, if the value of relative improvement of the number of passengers that finished their trip is smaller, then the approach has a better performance. For the other criteria in Table 6.8, a bigger value represents a better performance.

Table 6.7 Performance comparison of the all-stop approach, the bi-level approach, and the limited bi-level approach

Solution approach		Objective value (–)	Computation time (s)	Energy consumption (J)	Number of passengers		Travel time for passengers that (s)	Waiting time of passengers that (s)
					That finished their trips (passengers)	That did not travel (passengers)		
All-stop		2.917	4.473×10^2	5.888×10^9	2.603×10^4	1.324×10^4	1.944×10^7	1.031×10^7
Bi-level		2.617	1.671×10^4	4.994×10^9	2.451×10^4	1.476×10^4	1.648×10^7	1.270×10^7
Limited bi-level	Threshold	2.708	1.192×10^3	5.369×10^9	2.500×10^4	1.427×10^4	1.724×10^7	1.192×10^7
	$\chi_0 = 1$	2.681	1.698×10^3	5.301×10^9	2.505×10^4	1.422×10^4	1.706×10^7	1.188×10^7
	$\chi_0 = 2$	2.661	2.349×10^3	5.220×10^9	2.482×10^4	1.445×10^4	1.680×10^7	1.227×10^7
	$\chi_0 = 3$	2.653	4.574×10^3	5.111×10^9	2.473×10^4	1.454×10^4	1.683×10^7	1.240×10^7

These remaining passengers, i.e., the passengers that did not travel, will be picked up by the trains that arrive later on. The waiting time of those passengers is also included in the objective function. So here the number and the waiting time of these passengers are also given for comparing the performance of these three approaches

Table 6.8 Relative performance improvement of the bi-level approach and the limited bi-level approach with respect to the all-stop approach

Solution approach		Objective value (%)	Computation time	Energy consumption (%)	Number of passengers		Travel time for passengers that finished their trips (%)	Waiting time of passengers that did not travel (%)
					That finished their trips (%)	That did not travel (%)		
Bi-level		10.28	−36.36	15.18	5.83	−11.46	15.19	−23.26
Limited bi-level	Threshold	7.16	−1.66	8.81	3.94	−7.75	11.32	−15.64
	$\chi_0 = 1$	8.09	−2.80	9.97	3.77	−7.41	12.24	−15.29
	$\chi_0 = 2$	8.78	−4.25	11.35	4.64	−9.12	13.55	−19.04
	$\chi_0 = 3$	9.05	−9.23	13.20	5.01	−9.85	13.41	−20.35

6.5.2.2 Discussion

In this case study, the overall performance improvement of the stop-skipping strategy is about 8–12 % better when compared to the all-stop approach. With the stop-skipping strategy, the total travel time is reduced with 12–15 % and the total energy consumption is reduced with 10–15 %. The number of passengers who did not travel increases with 7–11 %; however, it should be noted that the trains coming later will pick up these passengers anyway; so those passengers will not be left at the platform. Since we solving the train scheduling problem in a rolling horizon way (cf. Sect. 7.4), the passengers who did not travel in the current time period will be taken into account in the next period. The limited bi-level approach yields an acceptable performance when compared with the bi-level approach. However, the computation time of the bi-level approach is about 10 and 4 times longer than that of the limited bi-level approach with $\chi_0 = 1$ and $\chi_0 = 3$, respectively. Note that the computation time of the approaches proposed in this chapter can be reduced by implementing the approaches in object code and by using faster processors and/or parallel processing.

6.6 Summary

We have considered the train scheduling problem with OD-dependent passenger demands for an urban rail transit line, where the operation of trains at both the stations and the terminal station are included in the model. Since the resulting train scheduling problem is a MINLP problem, a limited bi-level approach has been proposed, where a threshold method is applied to obtain a good initial solution for the full problem and where the search space for the variables can next be limited to enhance the efficiency. For a case study, the limited bi-level approach with a limited search space provides the best solution within the time that is typically available for the computations (e.g., half an hour). In particular, the overall performance improved with about 8–12 % compared to the all-stop approach.

In the future, we will investigate other solution approaches to solve the resulting MINLP problem efficiently, especially for cases with a large number of trains and stations, and we will compare these approaches with the threshold approach and the limited bi-level approach for large-scale real-life case studies. In addition, we will investigate the effect of more detailed models (including short turns, the distribution of on board passengers and waiting passengers at platforms, etc.) on the trade-off between performance and computational complexity. In future work, we will also consider the travel time uncertainty, dwell time uncertainty, etc., in the train scheduling and investigate robust train scheduling approaches.

References

1. Wang Y, De Schutter B, van den Boom T, Ning B, Tang T (2014) Efficient bi-level approach for urban rail transit operation with stop-skipping. IEEE Trans Intell Transp Syst 15:2658–2670
2. Wang Y, De Schutter B, van den Boom T, Ning B, Tang T (2014) Origin-destination dependent train scheduling problem with stop-skipping for urban rail transit systems. In: Proceedings of the 93rd annual meeting of the transportation research board, Washington, DC
3. Elberlein X, Wilson N, Bernstein D (2001) The holding problem with real-time information available. Transp Sci 35:1–18
4. Wang H, Schmid F, Chen L, Roberts C, Xu T (2013a) A topology-based model for railway train control systems. IEEE Trans Intell Transp Syst 14:819–827
5. Boggs P, Tolle J (1995) Sequential quadratic programming. Acta Numerica 4:1–51
6. Hooke R, Jeeves T (1961) "Direct search" solution of numerical and statistical problems. J Assoc Comput Mach 8:212–229
7. Gembicki F, Haimes Y (1975) Approach to performance and sensitivity multiobjective optimization the goal attainment method. IEEE Trans Autom Control 20:769–771
8. Kiresuk T, Sherman R (1968) Goal attainment scaling a general method for evaluating comprehensive community mental health programs. IEEE Trans Autom Control 4:443–453

Chapter 7
OD-Dependent Train Scheduling for an Urban Rail Transit Network

Abstract In the previous two chapters, we have discussed the train scheduling problem for a single urban rail transit line with OD-independent and OD-dependent passenger demands. In the current chapter, we consider the train scheduling problem for an urban rail transit *network*, i.e., a collection of interdependent transit lines. An event-driven model is built for train scheduling, which involves three types of events, i.e., departure events, arrival events, and passenger arrival rates change events. The routing of the arriving passengers at transfer stations is formulated in the train scheduling model. Furthermore, the passenger walking time and the passenger transfer process are also taken into account in the train scheduling model. The resulting optimization problem is a real-valued nonlinear nonconvex problem, which can be solved by nonlinear programming approaches and evolution algorithms. The performance and the effectiveness of the event-driven model are evaluated via a case study. This chapter is based on Wang et al. (Transp Res Part C 60: 1–23, 2015) [1].

7.1 Introduction

In Chaps. 5 and 6, we have studied the train scheduling problem for an urban rail transit line which takes the passenger demands into account. As in general urban rail transit lines are separated from each other, passengers may need to make several interchanges between different lines to arrive at their destination. Therefore, when scheduling trains for an urban rail network, it is important to take the passenger transfers into account to shorten the total travel time of passengers [1].

In order to model the time-varying passenger arrival rates, we propose an event-driven model for the train scheduling in this chapter. The event-driven model includes departure events, arrival events, and passenger arrival rate change events at platforms. At transfer stations, if there exist multiple route choices for passengers to arrive at their destinations, the arriving passengers will distribute themselves and go to different platforms of different lines. As for the passengers arriving at transfer stations by trains, some of them will get off trains and will transfer to other lines to arrive at their destinations. We introduce splitting rates to describe the route choice of passengers. The splitting rate is defined as a fraction of the total passenger flow with the same

Y. Wang et al., *Optimal Trajectory Planning and Train Scheduling for Urban Rail Transit Systems*, Advances in Industrial Control,
DOI 10.1007/978-3-319-30889-0_7

destination that goes to a specific platform. Both the changes of splitting rates and the passenger transfers result in passenger arrival rate change events at transfer stations. This behavior can be captured by the event-driven model. The resulting train scheduling problem is a real-valued nonlinear nonconvex programming problem.

The rest of the chapter is structured as follows. Section 7.2 introduces the three types of events and formulates the event-driven dynamics of train scheduling. Section 7.3 elaborates the performance criteria and constraints of the train scheduling problem. In Sect. 7.4, we discuss how to solve the scheduling problem in a rolling horizon way and how to define the initial conditions for the scheduling problem. Several solution approaches, e.g., the sequential quadratic programming (SQP) method and genetic algorithm, are introduced to solve the resulting nonlinear nonconvex optimization problem in Sect. 7.5. In Sect. 7.6, the performance of the proposed event-driven model is evaluated via a case study. Finally, conclusions and recommendations are provided in Sect. 7.7.

7.2 Model Formulation

For an urban rail transit network with L lines and J stations, we denote the sets of lines and station indices as S_{ln} and S_{sta}. In general, a station could have several platforms and we denote the set of platforms as S_{pla}. It should be noted that an urban rail transit line with two directions is considered as two separate lines in this chapter. We make the following assumptions:

A.1 There is no shared platform for different lines in the urban rail transit network. If passengers want to transfer from one line to the other, they need to walk from one platform to the other.

A.2 A platform can only accommodate one train at a time and no overtaking can occur at any point of the line.

Assumption A.1 holds for most urban transit systems, e.g., the subway networks in Beijing, Paris, and Rome. With Assumption A.1, a platform is uniquely identified to a specific line; so a line can be defined by a subset of platforms. In general, Assumption A.2 also holds for most urban transit systems. Moreover, the trains of different urban rail transit lines are operated separately in practice, which means that trains cannot be shared between different lines.

If platform p of station j is on line ℓ, we denote the predecessor of platform p on line ℓ as $\mathfrak{p}^{\text{pla}}(p)$ and the successor of platform p on line ℓ as $\mathfrak{s}^{\text{pla}}(p)$. Train service numbers are introduced to distinguish the different running cycles of the physical trains. So each train service in the network has a unique service number which uniquely identifies a train and its running cycle. After the arrival of a physical train at the terminal station, its service number will be augmented the total number of trains in the network when the train departs. Let I_ℓ be the total number of physical trains in transit line ℓ, then the total number of physical trains in the network is denoted as I_{net}, which can be calculated by $I_{\text{net}} = \sum_{\ell \in S_{\text{ln}}} I_\ell$. In addition, the set of indices of all train services is defined as $S_{\text{tra}} = \{1, 2, \ldots, I_{\text{net}}, I_{\text{net}} + 1, \ldots, 2I_{\text{net},\ldots,N_{\text{cyc}}I_{\text{net}}}\}$ with

$N_{\rm cyc}$ the number of running cycles for trains. The predecessor and successor of train service i on line ℓ are denoted as $\mathfrak{p}^{\rm tra}(i)$ and $\mathfrak{s}^{\rm tra}(i)$, respectively. The start time and end time of the scheduling period are denoted as t_0 and $t_{\rm end}$.

7.2.1 Three Types of Events

We model the train scheduling problem considering passenger demands by the following three types of events:

- Departure event: Representing the departure of a train at a station.
- Arrival event: Representing the arrival of a train at a station.
- λ-change event: Representing the change of passenger arrival rates at a platform.

In order to describe the operation of trains, we propose an event-driven model which consists of the discrete events listed above and a continuous part that describes the movement of a train running from one station to another station in the network. The k-th event e_k occurring in the event-driven system can be written as

$$e_k = (t_k, Y_{{\rm type},k}, i_k, p_k), \tag{7.1}$$

where k is the event counter, t_k is the time instant at which event e_k occurs, $Y_{{\rm type},k}$ is the type of event e_k (which can have three possible values, i.e., 'd', 'a', or 'λ' corresponding to a departure event, an arrival event, or a λ-change event), i_k is the train index, and p_k is the platform index.

In particular, λ-change events can be triggered by the change of passenger arrivals at stations, the change of splitting rates at transfer stations, and the passenger transfers at transfer stations. It can be noted that the passenger arrival rate does not change between any two subsequent events. The train scheduling model requires the real-time assessment of passenger arrival rates for different origins and destinations during the scheduling period. The passenger arrival rates are available in the case of full state information. However, the passenger arrival rates are unknown in practice, but need to be estimated using the historical data and the current passenger flows collected by the advanced fare collection systems [2]. The solid line in Fig. 7.1 shows a typical profile for passenger arrivals at stations on workdays. The passenger arrival rate during the peak hours is much higher than that during the off-peak hours. The continuous passenger arrival rate can be approximated by a piecewise constant function as indicated by the dashed line in Fig. 7.1. Piecewise constant function $\lambda_{j,m}^{\rm station}(\cdot)$ with $j, m \in S_{\rm sta}$ denote the passenger arrival rate at station j of passengers with station m as their final destination. These passenger arrival rates, described by piecewise constant functions, are the inputs to the event-driven model. We describe these piecewise constant functions via so-called base profiles, which are left-hand side continuous piecewise constant functions. Moreover, these functions can be specified by a list of corner points as shown in Fig. 7.2, where the corner points are marked

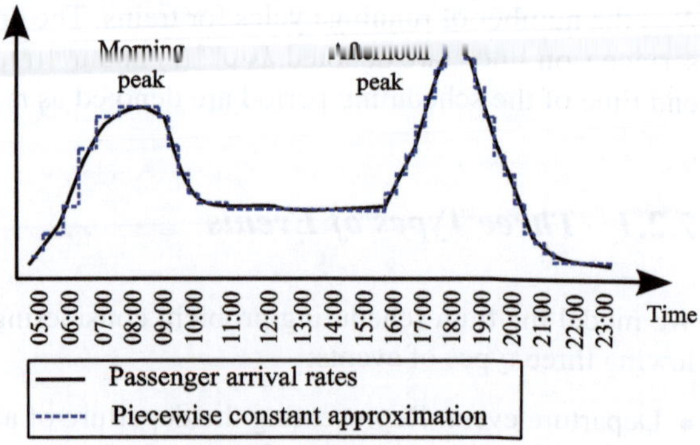

Fig. 7.1 Typical passenger arrival rate profile at a station for a working day

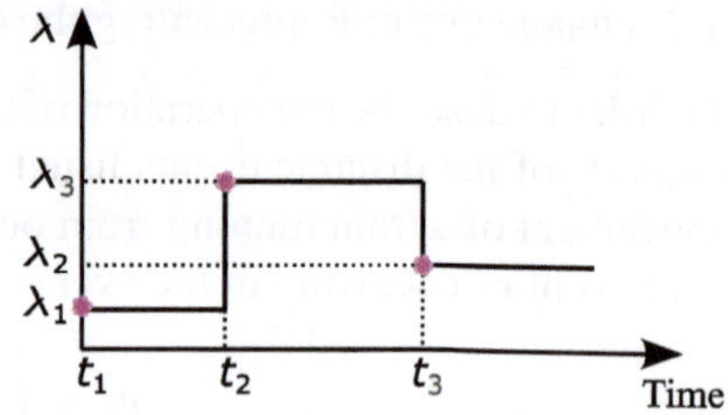

Fig. 7.2 Example of base profile with corner points for passenger arrival rates

with big dots. Therefore, the base profile shown in Fig. 7.2 can be described by these three corner points:

$$\{(t_1, \lambda_1), (t_2, \lambda_2), (t_3, \lambda_3)\}.$$

In particular, we introduce a λ-profile query module for each platform (see also Fig. 7.3) to obtain the passenger arrival rate for any time instant [3]. If platform p is not at a transfer station, then the query module only includes the base profile $\lambda^{\text{station}}_{j,m}(\cdot)$. However, if platform p belongs to a transfer station, then the query module for this platform contains the base profile and possibly additional update profiles caused by splitting rate changes and passenger transfers:

1. Splitting rates change at stations: At a transfer station, there could be multiple routes available for passengers to go to their destinations. So passengers with the same destination can choose to go to different platforms at a transfer station. Rail operators can influence or control the splitting of passenger flows at transfer stations by providing route information and suggestions to passengers through personal digital assistant (PDA) devices or through information panels at the entrance of stations. Let $\beta^{\text{station}}_{p,m}(\cdot)$ denote the splitting rate for the passengers flows that arrive at station j, have destination m, and go to platform p with $p \in P_j$ (see Fig. 7.4). Note that the function $\beta^{\text{station}}_{p,m}(\cdot)$ is also defined as a left-hand side continuous piecewise constant function. The splitting rate should not change too frequently, e.g., 15 min, to provide a consistent service to passengers. The passenger arrival rate at platform p of station j can be computed as follows:

$$\lambda_{p,m}(t_k) = \beta^{\text{station}}_{p,m}(t_k)\lambda^{\text{station}}_{j,m}(t_k), \quad \forall p \in P_j, \forall m \in S_{\text{sta}}, \qquad (7.2)$$

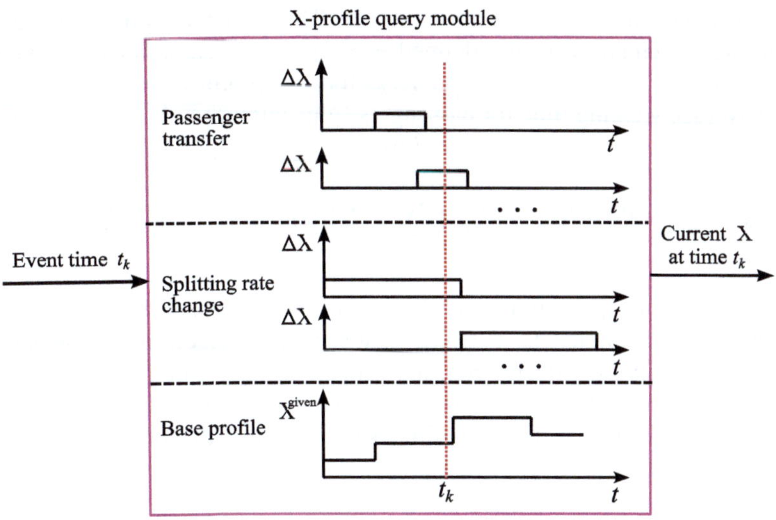

Fig. 7.3 λ-profile query module

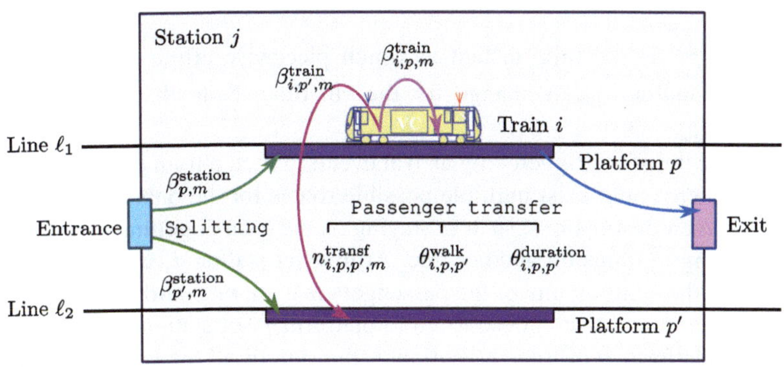

Fig. 7.4 Variables for the splitting of passenger flows and passenger transfers. **a** Variables immediately before the boarding process of passengers. **b** Variables immediately after the boarding process of passengers

where t_k is one of the corner points of the base profile or of the splitting rate change profiles and P_j is the set of platforms at transfer station j. Moreover, the sum of all the splitting rates at transfer station j is equal to 1, i.e.,

$$\sum_{p \in P_j} \beta^{station}_{p,m}(t_k) = 1, \quad \forall m \in S_{sta}, \ \forall j \in S^{transf}_{sta}, \tag{7.3}$$

where S^{transf}_{sta} is the set of transfer stations in the network. The splitting rates at the transfer stations are the control variables for the proposed train scheduling problem. The change of the splitting rates of passenger flows leads to λ-change

events at the platforms of a transfer station. The $\Delta\lambda$ update profiles triggered by splitting rate changes are also defined as piecewise constant functions, which can be described by a list of corner points as for base profiles.
The average walking time for passengers from entrances of station j to platform p at time instant t can be calculated by

$$\theta_p^{\text{walk-in}}(t) = a_{0,p}^{\text{walk}}\Big(\sum_m \beta_{p,m}^{\text{station}}(t)\lambda_{j,m}^{\text{station}}(t)\Big) + b_{0,p}^{\text{walk}}, \tag{7.4}$$

where $a_{0,p}^{\text{walk}}$ and $b_{0,p}^{\text{walk}}$ are the coefficients of the average walking time from the entrance of the station to platform p, which are decided by the layout of the station, the walking distance, etc., and which can, e.g., be determined on the basis of historical data. For all the passengers going from entrances of station j to platform p during the scheduling time period $[t_0, t_{\text{end}}]$, the total walking time can be computed by

$$t_p^{\text{walk-in}} = \sum_{c=1}^{C} \theta_p^{\text{walk-in}}(t_{p,c}^{\text{station}}) \sum_m \beta_{p,m}^{\text{station}}(t_{p,c}^{\text{station}})\lambda_{j,m}^{\text{station}}(t_{p,c}^{\text{station}})(t_{p,c+1}^{\text{station}} - t_{p,c}^{\text{station}}), \tag{7.5}$$

where $t_{p,c}^{\text{station}}$ is the time instant at which piecewise affine constant functions $\beta_{p,m}^{\text{station}}(\cdot)$ and/or $\lambda_{j,m}(\cdot)$ change for the cth time. Note that $t_{p,1}^{\text{station}} = t_0$ and $t_{p,C}^{\text{station}} = t_{\text{end}}$.

2. Passenger transfers triggered by arrival events: When a train arrives at a transfer station, there could exist multiple possible routes for the onboard passengers to arrive at their destinations, such as staying on the train or getting off the train and transferring to trains on another line. At transfer station j, we introduce $\beta_{i,p',m}^{\text{train}}$ to denote the splitting rate of the passengers that are on board of train i and have destination m and that choose to go to platform p' of station j. For train i that stops at platform p of transfer station j, the sum of all the splitting rates has to be equal to 1, i.e.,

$$\sum_{p' \in P_j} \beta_{i,p',m}^{\text{train}} = 1, \quad \forall i \in S_{\text{tra}}, \; \forall m \in S_{\text{sta}}. \tag{7.6}$$

It is noted that the passengers with destination j, i.e., the ones for which $m = j$, will not transfer to other platforms of station j but they will exit the transit network from station j; so if train i arrives at platform p, the value of $\beta_{i,p,j}^{\text{train}}$ is equal to 1 and the value of $\beta_{i,p',j}^{\text{train}}$ is equal to 0 for $p' \in P_j \setminus \{p\}$ where p is the platform of station j at which train i arrives. The total walking time for passengers to exit station j from platform p in the scheduling time period $[t_0, t_{\text{end}}]$ can be formulated as

$$t_p^{\text{walk-out}} = \sum_{i \in S_{p,t_0,t_{\text{end}}}^{\text{tra}}} a_{p,0}^{\text{walk}} n_{i,p,j}^{\text{alight}} + b_{p,0}^{\text{walk}}, \tag{7.7}$$

where $S^{\text{tra}}_{p,t_0,t_{\text{end}}}$ is the subset of indices of trains that stop at platform p in $[t_0, t_{\text{end}}]$, $n^{\text{alight}}_{i,p,j}$ is the number of passengers who alight from train i, have destination j, and exit the network from platform p. The coefficients $a^{\text{walk}}_{p,0}$ and $b^{\text{walk}}_{p,0}$ can be determined similarly as $a^{\text{walk}}_{0,p}$ and $b^{\text{walk}}_{0,p}$.

The transfer walking time for transfer passengers is up to the walking distance between two platforms and the number of transfer passengers. The walking time for transfer passengers could have the shape as shown by the line in Fig. 7.5 in practice. For the sake of simplicity, we approximate the relationship between the passenger walking time and number of transfer passengers by a piecewise constant function as illustrated by the dashed line in Fig. 7.5. So, the average walking time can be calculated for the transfer passengers from platform p to the other platforms $p' \in P_j \setminus \{p\}$ as follows:

$$\theta^{\text{walk}}_{i,p,p'} = a^{\text{walk}}_{p,p'} n^{\text{transf}}_{i,p,p'} + b^{\text{walk}}_{p,p'}, \quad \forall i \in S_{\text{tra}}, \ \forall p' \in P_j \setminus \{p\}, \tag{7.8}$$

where $n^{\text{transf}}_{i,p,p'}$ is the number of transfer passengers from train i to platform p' of line ℓ', $a^{\text{walk}}_{p,p'}$ and $b^{\text{walk}}_{p,p'}$ are the coefficients of the average walking time, which can, e.g., be determined by using historical data and which rely on the layout of transfer station, the walking distance, etc. The transfer time $t^{\text{transf}}_{i,p}$ for all transferring passengers getting off from train i can be calculated by

$$t^{\text{transf}}_{i,p} = \sum_{p' \in P_j \setminus \{p\}} \theta^{\text{walk}}_{i,p,p'} n^{\text{transf}}_{i,p,p'}. \tag{7.9}$$

The duration time for the transfer process can be approximated in a similar way as the average walking time by

$$\theta^{\text{duration}}_{i,p,p'} = a^{\text{duration}}_{p,p'} n^{\text{transf}}_{i,p,p'} + b^{\text{duration}}_{p,p'}, \quad \forall i \in S_{\text{tra}}, \ \forall p' \in P_j \setminus \{p\}, \tag{7.10}$$

where $a^{\text{duration}}_{p,p'}$ and $b^{\text{duration}}_{p,p'}$ can be determined based on historical data like $a^{\text{walk}}_{p,p'}$ and $b^{\text{walk}}_{p,p'}$. The updates for the λ-profile due to passenger transfers as given in Fig. 7.3 can also be described by a list of corner points

$$\left\{(a_{i,j}, 0), (a_{i,j} + \theta^{\text{walk}}_{i,p,p'}, \Delta\lambda_{i,p,p'}), (a_{i,j} + \theta^{\text{walk}}_{i,p,p'} + \theta^{\text{duration}}_{i,p,p'}, 0)\right\}, \tag{7.11}$$

where $\Delta\lambda_{i,p,p'}$ is calculated by

$$\Delta\lambda_{i,p,p'} = \frac{n^{\text{transf}}_{i,p,p'}}{\theta^{\text{duration}}_{i,p,p'}}. \tag{7.12}$$

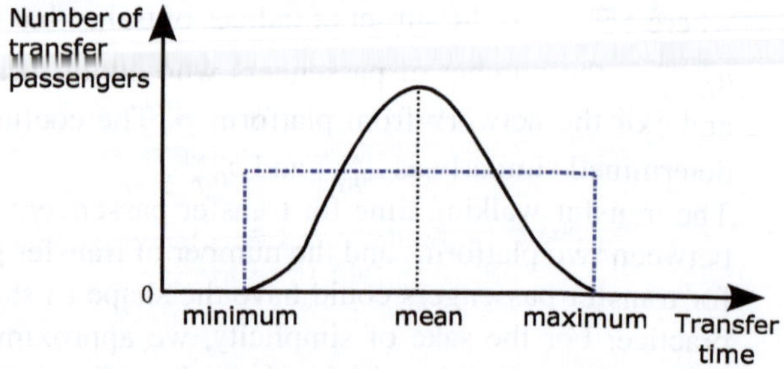

Fig. 7.5 Typical walking time profile for the transfer passengers

7.2.2 *Event-Driven Dynamics*

In this event-driven system, there are two classes of events: autonomous events and controlled events. Autonomous (or triggered) events are triggered by other events or by the environment, and their event times cannot be controlled directly. All the λ-change events are autonomous events. The departure and arrival events at stations are the controlled events of the given system. The event times of the controlled events are directly influenced by the inputs to the system. The control inputs of the system are the departure times, the running times of trains, and the splitting rates of passenger flows at transfer stations. The model structure of the event-driven system is illustrated in Fig. 7.6.

A global event list is introduced for the event-driven system (see also Fig. 7.6), which includes all the possible next events for all the trains and stations in the urban rail network. The next event of the system will be the event that will occur first, i.e., the event with the smallest value of t_k in the global event list. In the beginning, the global event list should be initialized according to the initial state of the system. We denote the current time as t_{current} (see also Fig. 7.7). Let τ_{process} be the processing time for data preparation of the train scheduling. All the events that happened in the past, i.e., for which the event time is smaller than $t_{\text{current}} + \tau_{\text{process}}$, are known to the event-driven system and the set of these events is denoted as S_{known}. The set of events that will happen in the future is denoted as S_{unknown}.

When an event happens, the state of the system should be updated and some other events may be triggered. For any event occurring in the event-driven system, we need to update the number of passengers waiting at platforms. It is important

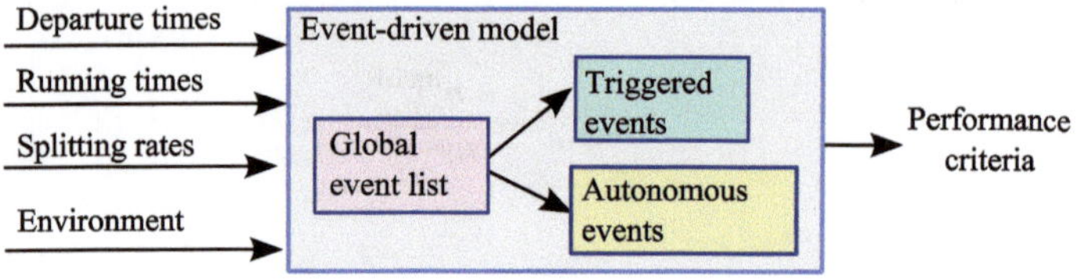

Fig. 7.6 Model structure of the event-driven system

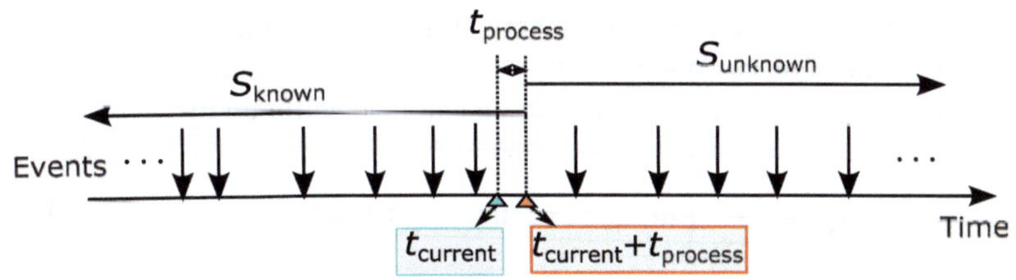

Fig. 7.7 Definition of known events and unknown events

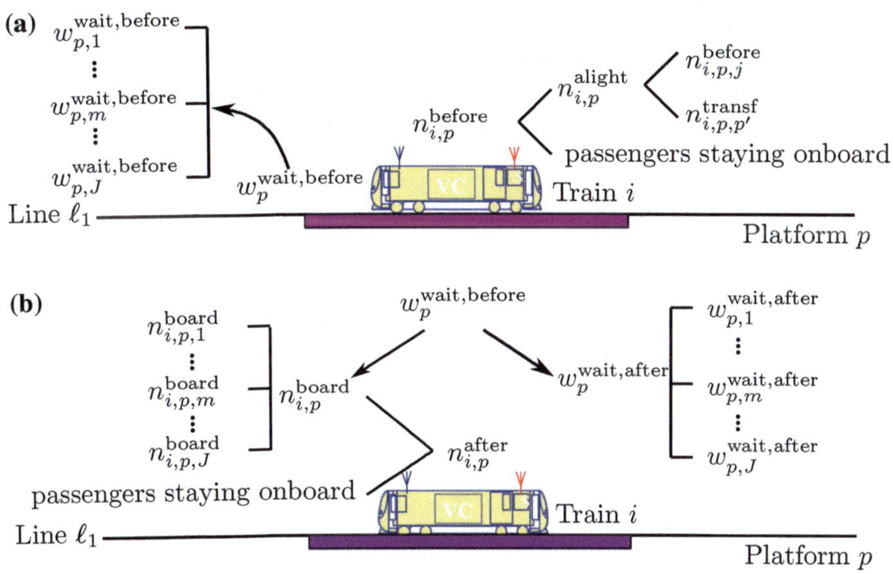

Fig. 7.8 Variables for the passenger characteristics before and after the boarding process. **a** The alighting process of passengers at station j, **b** the boarding process of passengers at station j

to note that the passenger arrival rate does not change between any two subsequent events. Immediately before event e_k happens, we calculate the number of passengers $w_{p_k,m}^{\text{wait,before}}(t_k)$ with destination m that are waiting at platform p_k as follows (see Fig. 7.8):

$$w_{p_k,m}^{\text{wait,before}}(t_k) = w_{p_k,m}^{\text{wait,after}}(t_{k'}) + \lambda_{p_k,m}(t_{k'})(t_k - t_{k'}), \tag{7.13}$$

where $t_{k'}$ is the event time of the previous event $e_{k'} = (t_{k'}, Y_{\text{type},k'}, i_{k'}, p_{k'})$ happening at platform p_k with $p_{k'} = p_k$, $w_{p_k,m}^{\text{wait,after}}(t_{k'})$ is the number of passengers at the platform immediately after event $e_{k'}$, and $\lambda_{p_k,m}(t_{k'})(t_k - t_{k'})$ is the number of passengers that arrive at platform p_k between t_k' and t_k. Moreover, we compute the total number of waiting passengers $w_{p_k}^{\text{wait,before}}(t_k)$ at platform p_k immediately before the event e_k by

$$w_{p_k}^{\text{wait,before}}(t_k) = \sum_{m \in S_{\text{sta}}} w_{p_k,m}^{\text{wait,before}}(t_k). \tag{7.14}$$

Furthermore, we also need to update the waiting time of passengers at a platform when an event takes place. When event e_k occurs, we denote the waiting time of the passengers at platform p_k as $t_{p_k}^{\text{wait}}(t_k)$, which can be calculated by

$$t_{p_k}^{\text{wait}}(t_k) = t_{p_k}^{\text{wait}}(t_k') + \sum_{m \in S_{\text{sta}}} \left(w_{p_k,m}^{\text{wait,after}}(t_{k'})(t_k - t_{k'}) + \frac{1}{2}\lambda_{p_k,m}(t_{k'})(t_k - t_{k'})^2 \right), \quad (7.15)$$

with $t_{k'}$ the event time of the previous event $e_{k'}$ that occurred at platform p_k.

Generally, the updates of other states and the triggered events caused by the current event rely on the type of the current event. For λ-change events, we only need to update the number of waiting passengers at platforms and the waiting time of these passengers. For departure events and arrival events, the updates of other states and the triggered events are explained in detail as follows.

7.2.2.1 Departure Events

When a departure event occurs, denoted as e_k, then we have $Y_{\text{type},k} =$ 'd'. Let i_k be the train involved in the event. The number of passengers boarding train i_k at platform p_k on line ℓ_k is equal to the minimum of the remaining space $n_{i_k,p_k}^{\text{remain}}$ on the train after the alighting process and the number of waiting passengers $w_{p_k}^{\text{wait,before}}(t_k)$, which can be written as

$$n_{i_k,p_k}^{\text{board}} = \min(n_{i_k,p_k}^{\text{remain}}, w_{p_k}^{\text{wait,before}}(t_k)). \quad (7.16)$$

The remaining space $n_{i_k,p_k}^{\text{remain}}$ on train i_k for passengers is

$$n_{i_k,p_k}^{\text{remain}} = C_{\max,i_k} - n_{i_k,\mathfrak{p}^{\text{pla}}(p_k)} - n_{i_k,p_k}^{\text{alight}}, \quad (7.17)$$

where $C_{\max,i_k}$ is the maximum capacity of train i_k, $\mathfrak{p}^{\text{pla}}(p_k)$ is the predecessor platform[1] of platform p_k, $n_{\alpha,\beta}$ is the number of passengers on train α when it departs from platform β, and $n_{\alpha,\beta}^{\text{alight}}$ is the number of passengers getting off train α at platform β. The calculation for $n_{i_k,p_k}^{\text{alight}}$ will be given in (7.30).

The number of passengers $w_{p_k}^{\text{wait,after}}(t_k)$ left by train i_k at platform p_k, i.e., the number of passengers waiting at the platform immediately after event e_k, can be computed by

$$w_{p_k}^{\text{wait,after}}(t_k) = w_{p_k}^{\text{wait,before}}(t_k) - n_{i_k,p_k}^{\text{board}}. \quad (7.18)$$

Here, we assume that the number of passengers with destination m that are left by train i_k is proportional to the number of waiting passengers. Hence, the number of passengers with destination m left by train i_k at platform p_k can be calculated as

[1] Recall that $\mathfrak{p}^{\text{pla}}(p_k)$ is the previous platform on the line to which platform p_k belongs.

$$w_{p_k,m}^{\text{wait,after}}(t_k) = w_{p_k}^{\text{wait,after}}(t_k)\frac{w_{p_k,m}^{\text{wait,before}}(t_k)}{w_{p_k}^{\text{wait,before}}(t_k)}. \tag{7.19}$$

So the number of passengers with destination m that board train i_k at platform p_k can be written as

$$n_{i_k,p_k,m}^{\text{board}} = w_{p_k,m}^{\text{wait,before}}(t_k) - w_{p_k,m}^{\text{wait,after}}(t_k). \tag{7.20}$$

After the boarding process, the number of passengers $n_{i_k,p_k,m}^{\text{after}}$ with destination m that are on board of train i_k is updated as

$$n_{i_k,p_k,m}^{\text{after}} = n_{i_k,p_k,m}^{\text{before}} + n_{i_k,p_k,m}^{\text{board}}, \tag{7.21}$$

where $n_{i_k,p_k,m}^{\text{before}}$ is the number of passengers that are on board train i_k and have destination m before the boarding process of passengers. In addition, the number of passengers $n_{i_k,p_k}^{\text{after}}$ on board of train i_k at platform p_k after the boarding process can be calculated as

$$n_{i_k,p_k}^{\text{after}} = n_{i_k,p_k}^{\text{before}} + n_{i_k,p_k}^{\text{board}}, \tag{7.22}$$

where $n_{i_k,p_k}^{\text{before}}$ is the total number of passengers on board train i_k before the boarding process of passengers (see more details in the description of arrival events).

The departure event e_k at platform p_k will generate an arrival event at the next platform of the line to which platform p_k belongs, which can be written as follows:

$$\left(a_{i_k,\mathfrak{s}^{\text{pla}}(p_k)}, \text{'a'}, i_k, \mathfrak{s}^{\text{pla}}(p_k)\right),$$

where $a_{i_k,\mathfrak{s}^{\text{pla}}(p_k)}$ is the arrival time of train i_k at platform $\mathfrak{s}^{\text{pla}}(p_k)$. The arrival time $a_{i_k,\mathfrak{s}^{\text{pla}}(p_k)}$ can be calculated by

$$a_{i_k,\mathfrak{s}^{\text{pla}}(p_k)} = d_{i_k,p_k} + r_{i_k,p_k}, \tag{7.23}$$

where d_{i_k,j_k} is equal to t_k and r_{i_k,p_k} is the running time on the track segment between platform p_k and platform $\mathfrak{s}^{\text{pla}}(p_k)$. This arrival event should then be added to the global event list.

7.2.2.2 Arrival Events

If $Y_{\text{type},k} =$ 'a', i.e., event e_k is the arrival event of train i_k at platform p_k on ℓ_k, the number of passengers $w_{p_k}^{\text{wait,before}}(t_k)$ waiting at platform p_k immediately before this arrival event should be updated by (7.13) and (7.14).

The number of passengers getting off train i_k depends on platform p_k:

- If platform p_k is at the first station of line ℓ_k, no passengers gets off train i_k, i.e.,

$$n_{i_k,p_k}^{\text{alight}} = 0. \tag{7.24}$$

Moreover, immediately before the boarding process at platform p_k the number of passengers $n^{\text{before}}_{i_k,p_k,m}$ that have destination m and are on board of train i_k is also equal to zero, i.e.,

$$n^{\text{before}}_{i_k,p_k,m} = 0, \quad \forall m \in S_{\text{sta}}. \tag{7.25}$$

- If the station j_k to which platform p_k belongs, is neither the first station nor a transfer station, then the passengers with destination j_k will get off train i_k. The number of alighting passengers can be calculated as follows:

$$n^{\text{alight}}_{i_k,p_k} = n^{\text{after}}_{i_k,\mathfrak{p}^{\text{pla}}(p_k),j_k}, \tag{7.26}$$

where $n^{\text{after}}_{i_k,\mathfrak{p}^{\text{pla}}(p_k),j_k}$ is the number of onboard passengers with destination j_k after the boarding process at predecessor platform $\mathfrak{p}^{\text{pla}}(p_k)$. Moreover, the number of passengers $n^{\text{before}}_{i_k,p_k,m}$ on board train i_k immediately before the boarding process can be calculated as follows:

$$n^{\text{before}}_{i_k,p_k,m} = n^{\text{after}}_{i_k,\mathfrak{p}^{\text{pla}}(p_k),m}, \quad \forall m \in S_{\text{sta}} \setminus \{j_k\}. \tag{7.27}$$

Hence, the total number of passengers on board train i_k before the boarding process is written as

$$n^{\text{before}}_{i_k,p_k} = \sum_{m \in S_{\text{sta}} \setminus \{j_k\}} n^{\text{before}}_{i_k,p_k,m}. \tag{7.28}$$

- If the station j_k to which platform p_k belongs is a transfer station, then not only the passengers with destination j_k will get off train i_k, but the passengers with other destinations may also get off train i_k. The splitting rates for the passengers with destination m going to platform p' are denoted as $\beta^{\text{train}}_{i_k,p',m}$ for $p' \in P_{j_k}$. The number of passengers $n^{\text{before}}_{i_k,p_k,m}$ that have destination m and are on board of train i_k immediately before the boarding process can be calculated by

$$n^{\text{before}}_{i_k,p_k,m} = \beta^{\text{train}}_{i_k,p_k,m} n^{\text{after}}_{i_k,\mathfrak{p}^{\text{pla}}(p_k),m}, \quad \forall m \in S_{\text{sta}}, \tag{7.29}$$

where $n^{\text{after}}_{i_k,\mathfrak{p}^{\text{pla}}(p_k),m}$ is the number of onboard passengers with destination m immediately after the boarding process at predecessor platform $\mathfrak{p}^{\text{pla}}(p_k)$.
As mentioned in Sect. 7.2.1, the splitting rate $\beta^{\text{train}}_{p_k,i_k,j_k}$ equals 1 for the passenger flow with destination j_k. All these passengers will get off the train and exit the network from station j_k. For the passengers with destination m with $m \neq j_k$, the passengers staying on line ℓ_k will stay on train i_k. Therefore, the number of alighting passengers can be calculated by

$$n^{\text{alight}}_{i_k,p_k} = n^{\text{after}}_{i_k,\mathfrak{p}^{\text{pla}}(p_k)} - \sum_{m \in S_{\text{sta}} \setminus \{j_k\}} n^{\text{before}}_{i_k,p_k,m}, \tag{7.30}$$

where $n^{\text{after}}_{i_k,\mathrm{p}^{\text{pla}}(p_k)}$ in fact equals the number of passengers on board of train i_k when it arrives at platform p_k and $\sum_{m \in S_{\text{sta}} \setminus \{j_k\}} n^{\text{before}}_{i_k,j_k,m}$ is the total number of passengers staying on train i_k after the alighting process. The number of transferring passengers $n^{\text{transf}}_{i_k,p_k,p',m}$ that have destination m and transfer from platform p_k to some other platform p' can be calculated by

$$n^{\text{transf}}_{i_k,p_k,p',m} = \beta^{\text{train}}_{i_k,p',m} n^{\text{after}}_{i_k,\mathrm{p}^{\text{pla}}(p_k),m}, \quad \forall p' \in P_{j_k} \setminus \{p_k\}. \tag{7.31}$$

The total number of transfer passengers from train i_k is calculated by

$$n^{\text{transf}}_{i_k,p_k} = \sum_{p' \in P_{j_k} \setminus \{p_k\}} \sum_{m \in S_{\text{sta}} \setminus \{j_k\}} n^{\text{transf}}_{i_k,p_k,p',m}. \tag{7.32}$$

When arrival event e_k happens, the passenger in-vehicle time $t^{\text{in-vehicle}}_{i_k,p_k}$ should be updated for train i_k, which consists of the running time of train i_k from platform $\mathrm{p}^{\text{pla}}(p_k)$ to platform p_k and the dwell time at platform p_k. The passenger in-vehicle time can be calculated by

$$t^{\text{in-vehicle}}_{i_k,p_k} = n^{\text{after}}_{i_k,\mathrm{p}^{\text{pla}}(p_k)} r_{i_k,p_k} + (n^{\text{after}}_{i_k,\mathrm{p}^{\text{pla}}(p_k)} - n^{\text{alight}}_{i_k,p_k})(d_{i_k,p_k} - a_{i_k,p_k}), \tag{7.33}$$

where r_{i_k,p_k}, d_{i_k,p_k}, a_{i_k,p_k} are the running time, departure time, and arrival time of train i_k at platform p_k.

Moreover, when arrival event e_k occurs at a transfer station, the passenger transfer process will trigger the λ-change events, which can be written as

$$\left(a_{i_k,p_k} + \theta^{\text{walk}}_{i_k,p_k,p'}, \text{`}\lambda\text{'}, -, p_k\right), \quad \forall p' \in P_{j_k} \setminus \{p_k\}, \tag{7.34}$$

$$\left(a_{i_k,p_k} + \theta^{\text{walk}}_{i_k,p_k,p'} + \theta^{\text{duration}}_{i_k,p_k,p'}, \text{`}\lambda\text{'}, -, p_k\right), \quad \forall p' \in P_{j_k} \setminus \{p_k\}, \tag{7.35}$$

where '−' is a dummy place holder as there is no train included in these events. The above λ-change events should be added to the global event list.

7.3 Mathematical Formulation for the Scheduling Problem

7.3.1 Performance Criteria

In this chapter, we minimize the total travel time of all passengers and the energy consumption of the trains using a weighted sum strategy. Note that we can accommodate other performance criteria all well.

The total energy consumption for all I trains running with J stations can then be formulated as

$$E_{\text{total}} = \sum_{i\in S_{\text{tra}}} \sum_{p\in S_{\text{pla}}} (E_{i,p}^{\text{acc}} + E_{i,p}^{\text{hold}}), \tag{7.36}$$

where $E_{i,p}^{\text{acc}}$ and $E_{i,p}^{\text{hold}}$ can be calculated by (5.17) and (5.18). The total travel time of all passengers includes the passenger waiting time, the passenger in-vehicle time, and the passenger transfer time, which can then be formulated as

$$\begin{aligned} t_{\text{total}} = & \sum_{p\in S_{\text{pla}}} t_p^{\text{wait}} + \sum_{i\in S_{\text{tra}}} \sum_{p\in S_{\text{pla}}} t_{i,p}^{\text{in-vehicle}} + \sum_{i\in S_{\text{tra}}} \sum_{p\in S_{\text{pla}}} t_{i,p}^{\text{transf}} \\ & + \sum_{p\in S_{\text{pla}}} t_p^{\text{walk-in}} + \sum_{p\in S_{\text{pla}}} t_p^{\text{walk-out}}. \end{aligned} \tag{7.37}$$

Similar as the performance criteria in Sect. 5.3, a weighted sum strategy can be applied to solve the multi-objective optimization of the train scheduling problem, i.e., we consider

$$f_{\text{opt}} = \frac{E_{\text{total}}}{E_{\text{total,nom}}} + \lambda \frac{t_{\text{total}}}{t_{\text{total,nom}}}, \tag{7.38}$$

where λ is a nonnegative weight, and the normalization factors $E_{\text{total,nom}}$ and $t_{\text{total,nom}}$ are the nominal values of the total energy consumption and the total travel time of passengers, respectively. These nominal values can, e.g., be determined by a feasible initial schedule. As mentioned in Sect. 6.3, a penalty term of the waiting time of the passengers left by the last train during the scheduling period can also be introduced to spread trains over the entire scheduling period.

7.3.2 Constraints

The event times of all the events in the future event set S_{unknown} should satisfy

$$t_k \geq t_{\text{current}} + \tau_{\text{process}}, \quad \forall e_k \in S_{\text{unknown}}. \tag{7.39}$$

In addition, the event times of the departure events and the arrival events should satisfy the operational constraints as follows. The event time t_k for arrival event e_k, i.e., the arrival time a_{i_k,p_k} of train i_k at platform p_k, should satisfy the headway constraints:

$$a_{i_k,p_k} - d_{\text{p}^{\text{tra}}(p_k),p_k} \geq h_{p_k,\min}, \tag{7.40}$$

where $d_{\text{p}^{\text{tra}}(p_k),p_k}$ is the departure time of the previous train at platform p_k and $h_{p_k,\min}$ is the minimum headway at platform p_k to ensure the safe operation of trains. Moreover, the event time of the departure event, i.e., the departure time d_{i_k,p_k} of train i_k at platform p_k, should satisfy

$$d_{i_k,p_k} \geq a_{i_k,p_k} + \tau_{i_k,p_k,\min}, \tag{7.41}$$

$$d_{i_k,p_k} \leq a_{i_k,p_k} + \tau_{i_k,p_k,\max}, \tag{7.42}$$

where $\tau_{i_k,p_k,\min}$ and $\tau_{i_k,p_k,\max}$ are the minimal and maximal dwell time for train i_k at platform p_k. The minimal dwell time depends on the number of passengers getting off and getting on a train, which can be calculated as

$$\tau_{i_k,p_k,\min} = \min\left(\tilde{\tau}_{\min}, \alpha_{1,\mathrm{d}} + \alpha_{2,\mathrm{d}} n_{i_k,p_k}^{\mathrm{alight}} + \alpha_{3,\mathrm{d}} n_{i_k,p_k}^{\mathrm{board}} + \alpha_{4,\mathrm{d}} \left(\frac{w_{p_k}^{\mathrm{wait}}(t_k)}{n^{\mathrm{door}}}\right)^3 n_{i_k,p_k}^{\mathrm{board}}\right), \tag{7.43}$$

where $\tilde{\tau}_{\min}$ is the minimum dwell time predefined by railway operator, $\alpha_{1,\mathrm{d}}$, $\alpha_{2,\mathrm{d}}$, $\alpha_{3,\mathrm{d}}$, and $\alpha_{4,\mathrm{d}}$ are coefficients that can, e.g., be estimated according to historical data, n^{door} is the number of doors that are used for the boarding and alighting process, and $w_{i_k,p_k}^{\mathrm{wait}}/n^{\mathrm{door}}$ is the approximated number of passengers waiting at each door. Furthermore, the departure time d_{i_k,p_k} should satisfy the headway constraint as follows:

$$d_{i_k,p_k} - d_{p^{\mathrm{tra}}(p_k),p_k} \leq h_{p_k,\max}, \tag{7.44}$$

where $h_{p_k,\max}$ is the maximum departure-departure headway between trains at platform p_k to guarantee the passenger satisfaction.

Note that the running time $r_{i,p}$ should satisfy

$$r_{i_k,p_k,\min} \leq r_{i_k,p_k} \leq r_{i_k,p_k,\max}, \tag{7.45}$$

where $r_{i_k,p_k,\min}$ and $r_{i_k,p_k,\max}$ are the minimal and maximal running time of train i_k between platform p_k and successor platform $\mathfrak{s}^{\mathrm{pla}}(p_k)$, respectively. The minimum running time is restricted by the train characteristics and the condition of the line. The maximum running time is introduced to ensure the passenger satisfaction since if trains run too slow, the passengers may complain.

7.4 Rolling Horizon Approach and Initial Conditions

A rolling horizon approach can be adopted to solve the train scheduling problem. In this section, we will discuss the rolling horizon approach in detail and we define the initial conditions for the scheduling problem.

Since passenger demands vary with the time in a daily operation, the train scheduling problem can be solved in a rolling horizon way, by solving the scheduling problem, e.g., every half an hour, so as to adapt the train schedule to passenger demands in real time. This works as follows. First, the train scheduling problem is solved for some period $[t_0, t_{\mathrm{end}}]$ and the trains will be operated according to the resulting optimal schedule. After some period of time t_{p}, e.g., half an hour, we will run the optimization process again, but now for the period $[t_0+t_p, t_{end}+t_p]$ using the known, measured, or estimated states of the system at time t_0+t_p. Once the new optimal schedule is computed, it is executed for t_p time units, and next the whole process is repeated again for the period $[t_0+2t_p, t_{end}+2t_p]$ and so on, until the end of the daily operation of the urban rail transit system.

When solving the train scheduling problem in a rolling horizon way, some of the variables[2] will no longer be free variables but will have fixed, known values. Assuming that t_0 is the start time instant of the scheduling period, we now discuss the fixed variables for a line in an urban rail network:

- If train i is in the terminal station at time t_0, i.e., the arrival time $a_{i-I_{net},0}$ of train $i-I_{net}$ at the terminal station will be a known value with $a_{i-I_{net},0} < t_0$. So $a_{i-I_{net},0}$ is no longer an unknown variable.
- If train i is at a platform of a station at time t_0, we use p_{i,t_0} to denote that platform. The arrival time $a_{i,p_{i,t_0}}$ of train i at platform p_{i,t_0} is known. In addition, the departure times, the arrival times, and the running times before platform p_{i,t_0} are also known.
- If train i is running on a segment at t_0, we use p_{i,t_0} to denote the segment at which train i is running on at t_0. The departure time $d_{i,p_{i,t_0}}$ of train i at platform p_{i,t_0} is a known time value with$d_{i,p_{i,t_0}} < t_0$. In addition, all the departure times, arrival times, and running times before segment p_{i,t_0} are known. Furthermore, the running time $r_{i,p_{i,t_0}}$ on segment p_{i,t_0} is also fixed since we assume that the schedule of a train can only be changed at platforms. Therefore, the arrival time of train i at platform $\mathfrak{p}^{pla}(p_{i,t_0})$ is also known.

The number of passengers on the train and the number of passengers waiting at the platform are also known at time t_0.

7.5 Solution Approaches

The train scheduling problem for an urban rail transit network is a nonlinear nonconvex programming problem with objective function (7.38) and constraints (7.39)–(7.45). The train scheduling problem in Chap. 5 is also a nonlinear non-convex programming problem, where several approaches are proposed, such as the gradient-based SQP approach, the gradient-free pattern search method, mixed integer (non)linear programming (MINLP) approach, and the iterative convex programming (ICP) approach. These approaches can also be applied to the train scheduling problem for an urban rail transit network in this chapter. In addition, the evolutionary algorithms, such as genetic algorithms, can also be applied to this train scheduling problem [4–6].

---

[2] When the stop-skipping strategy is included in the train scheduling, the stopping variables should be fixed for trains that are already on their way in order to make sure all the passengers on these trains can arrive at their destinations.

7.6 Case Study

7.6.1 Setup

In order to illustrate the event-driven model for urban rail transit networks proposed in this chapter, a small network with two cyclic lines as shown in Fig. 7.9 is considered as a test case study. Line 1, i.e., the solid line in Fig. 7.9, has 1 terminal (station 1) and 5 normal stations and line 2, i.e., the dashed line in Fig. 7.9, has 1 terminal (station 7) and 7 normal stations. The line data of these two lines are given in Table 7.1, where the calculation of the minimum running time is the same as that presented in Sect. 6.5. For each cyclic line, there are 5 physical trains and the number of train services considered in the train scheduling problem is taken as 7. The parameters of trains and passengers are chosen as in Table 7.2. The passenger arrival rates at stations are given in Table 7.3, where the passenger arrival rates are piecewise constant functions that can change with time in a scheduling period and the passenger arrival rates at terminals, i.e., station 1 and station 7, are equal to 0. In addition, since we only consider one direction of the cyclic lines, no passenger is arriving at the last stations of these two lines, i.e., station 6 and station 12.

At time t_0 (chosen as 2500 s), the initial states of trains for line 1 are as follows: train 1 and train 2 are running from station 4 and 2 to station 5 and station 3, respectively, and their arrival times are fixed, at 2530 s and 2550 s respectively. The number of passengers on train 1 and 2 at time t_0 is given in Table 7.4 and the number of passengers waiting at the platforms of line 1 is shown in Table 7.5. For line 2, the initial states at time t_0 are as follows: train 21, 22, and 23 are running from station 11, 9, and 8 to

Fig. 7.9 Layout of a small urban rail transit network

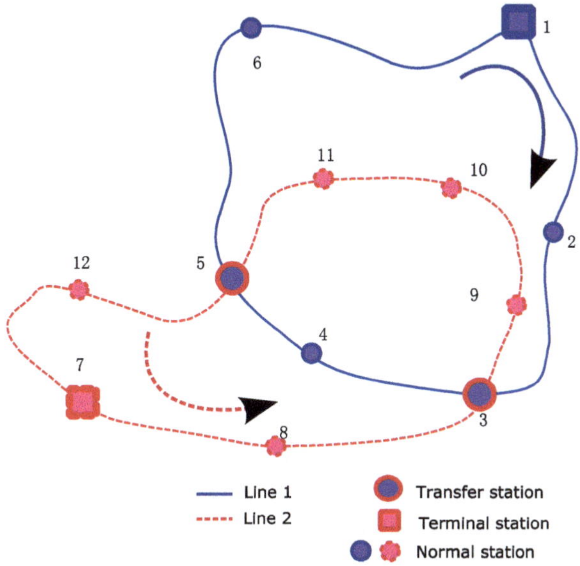

Table 7.1 Information of the two cyclic lines

Station number (Line 1)	1	2	3	4	5	6		
Distance to next station (m)	700	1500	1700	2200	1900	800		
Minimal running time (s)	59.278	95.282	104.283	126.785	113.284	63.779		
Station number (Line 2)	7	8	3	9	10	11	5	12
Distance to next station (m)	860	1400	1500	1300	1600	1200	1100	730
Minimal running time (s)	66.479	90.781	95.282	86.281	99.782	81.780	77.280	66.628

Table 7.2 Parameters of the trains and the passengers [1]

Property	Symbol	Value
Train mass (kg)	$m_{e,i}$	1.99×10^5
Mass of one passenger (kg)	m_p	60
Capacity of trains (passengers)	$C_{i,\max}$	1.50×10^3
Minimum dwell time (s)	$\tau_{\min}$	30
Maximum dwell time (s)	$\tau_{\max}$	150
Coefficients of the minimal dwell time		
(s)	$\alpha_{1,d}$	4.00
(s/passengers)	$\alpha_{2,d}$	4.7×10^{-2}
(s/passengers)	$\alpha_{3,d}$	5.1×10^{-2}
(s/passengers^{-4})	$\alpha_{4,d}$	1.00×10^{-6}
Coefficients of resistance		
(m/s^2)	k_{1i}	1.22×10^{-2}
(s^{-1})	k_{2i}	5.05×10^{-4}
(m^{-1})	k_{3i}	2.05×10^{-5}

station 5, 10, and 3, respectively and their arrival times are 2520, 2540, and 2560 s. The number of passengers on these trains at t_0 is given in Table 7.4 and the number of passengers waiting at platforms of line 2 is shown in Table 7.5. Furthermore, there are 3 and 2 trains stopping at terminal stations 1 and 7, respectively. We choose

Table 7.3 Passenger arrival rates (passengers/s) in the small urban rail network [1]

Station	Time period (s)	1	2	3	4	5	6	7	8	9	10	11	12
1	2500–5000	0	0	0	0	0	0	0	0	0	0	0	0
2	2500–3000	0	0	0.48	0.64	0.32	0.32	0	0	0.64	0.48	0.32	0.32
	3000–3600	0	0	0.32	0.48	0.32	0.16	0	0	0.48	0.40	0.35	0.16
	3600–5000	0	0	0.32	0.32	0.32	0.32	0	0	0.48	0.38	0.56	0.16
3	2500–3100	0	0	0	0.32	0.32	0.16	0	0	0.32	0.34	0.29	0.32
	3100–3700	0	0	0	0.48	0.32	0.32	0	0	0.32	0.22	0.42	0.32
	3700–5000	0	0	0	0.16	0.64	0.16	0	0	0.48	0.38	0.26	0.16
4	2500–3250	0	0	0	0	0.45	0.30	0	0	0	0	0	0.30
	3250–5000	0	0	0	0	0.53	0.38	0	0	0	0	0	0.38
5	2500–2850	0	0	0	0	0	0.60	0	0	0	0	0	0.30
	2850–3390	0	0	0	0	0	0.75	0	0	0	0	0	0.30
	3390–3830	0	0	0	0	0	0.60	0	0	0	0	0	0.30
	3830–5000	0	0	0	0	0	0.75	0	0	0	0	0	0.30
6	2500–5000	0	0	0	0	0	0	0	0	0	0	0	0
7	2500–5000	0	0	0	0	0	0	0	0	0	0	0	0
8	2500–3100	0	0	0.12	0.24	0.36	0.12	0	0	0.24	0.29	0.22	0.24
	3100–3700	0	0	0.12	0.24	0.60	0.12	0	0	0.24	0.34	0.19	0.24
	3700–5000	0	0	0.12	0.24	0.36	0.12	0	0	0.24	0.29	0.29	0.24
9	2500–3100	0	0	0	0	0.12	0.24	0	0	0	0.14	0.19	0.24
	3100–5000	0	0	0	0	0.12	0.36	0	0	0	0.29	0.31	0.36
10	2500–2900	0	0	0	0	0.15	0.15	0	0	0	0	0.21	0.15
	2900–5000	0	0	0	0	0.15	0.15	0	0	0	0	0.24	0.21
11	2500–2900	0	0	0	0	0.24	0.20	0	0	0	0	0	0.24
	2900–5000	0	0	0	0	0.36	0.32	0	0	0	0	0	0.28
12	2500–5000	0	0	0	0	0	0	0	0	0	0	0	0

the length of the considered scheduling period as 2500 s, where the schedule of 7 train services is optimized for each line. To ensure the passenger satisfaction, the maximum departure-departure headway at stations is taken as 400 s. The nominal values for the total travel time, the energy consumption, and the waiting time for the passengers who did not travel in the scheduling period is calculated according to a feasible schedule, which are 1.454×10^7 s, 3.436×10^9 J, and 7.434×10^6 s respectively.

The model formulation in Sect. 7.2 distinguishes the splitting rates $\beta_{p,m}^{\text{station}}$ of the passengers just entering the rail network and the splitting rates $\beta_{i,p,m}^{\text{train}}$ of the passenger arriving the transfer stations by trains. For this case study, we make[3] $\beta_{i,p,m}^{\text{train}}$ equal to

[3]Recall that the splitting rates $\beta_{i,p,m}^{\text{train}}$ for trains are numbers and that the splitting rates $\beta_{p,m}^{\text{station}}$ are piecewise constant functions that have the time as their argument.

Table 7.4 Number of passengers on board of trains at time t_0 for the two cyclic lines

Destination station	1	2	3	4	5	6	7	8	9	10	11	12	Total number of passengers
Train 1 (Line 1)	0	0	0	0	125	153	0	0	0	0	0	84	362
Train 2 (Line 1)	0	0	84	70	92	50	0	0	63	134	128	81	702
Train 6 (Line 2)	0	0	0	0	234	282	0	0	0	0	0	165	681
Train 7 (Line 2)	0	0	0	0	160	180	0	0	0	120	134	162	756
Train 8 (Line 2)	0	0	79	98	100	130	0	0	120	80	120	80	787

Table 7.5 Number of waiting passengers at platforms of the two cyclic lines

Destination	1	2	3	4	5	6	7	8	9	10	11	12	Total number of passengers
Station 1	0	0	0	0	0	0	0	0	0	0	0	0	0
Station 2	0	0	120	240	143	10	0	0	82	196	250	138	1179
Station 3	0	0	0	150	203	128	0	0	0	0	0	90	571
Station 4	0	0	0	0	210	230	0	0	0	0	0	112	552
Station 5	0	0	0	0	0	212	0	0	0	0	0	0	212
Station 6	0	0	0	0	0	0	0	0	0	0	0	0	0
Station 7	0	0	0	0	0	0	0	0	0	0	0	0	0
Station 8	0	0	150	120	80	103	0	0	180	100	140	140	1013
Station 3	0	0	0	0	100	128	0	0	190	110	130	100	758
Station 9	0	0	0	0	110	151	0	0	0	100	160	120	641
Station 10	0	0	0	0	130	170	0	0	0	0	130	150	580
Station 11	0	0	0	0	98	186	0	0	0	0	0	175	459
Station 5	0	0	0	0	0	0	0	0	0	0	0	212	212
Station 12	0	0	0	0	0	0	0	0	0	0	0	0	0

$\beta_{p,m}^{\text{station}}(a_{i,j})$ with $a_{i,j}$ the arrival time of train i at station j to simplify the train scheduling model. After this simplification, the number of decision variables of the problem can be reduced significantly, especially for cases with a large number of trains. Moreover, the walk-in and walk-out time for passengers in (7.37) are also considered as zero for the sake of simplicity. Two solution approaches are proposed to solve the train scheduling problem, i.e., the multi-start SQP approach and the genetic algorithm. The SQP method implemented by the fmincon function of MATLAB® optimization toolbox is employed and five feasible initial points are used to solve the optimization problem. For the genetic algorithm, the ga function of the global optimization toolbox of MATLAB is employed.

7.6.2 Results and Discussion

The train schedules obtained by the SQP method for line 1 and line 2 are shown in Figs. 7.10 and 7.11. The number of passengers on board of trains is shown in Figs. 7.12 and 7.13. Since there are no passenger arrivals at terminal stations (station 1 and station 7 for line 1 and line 2), the number of passengers on board of trains should be equal to 0 when trains depart or arrive at terminal stations, which is illustrated in Figs. 7.12 and 7.13. In Fig. 7.12, it is shown that when train 3 departs from station 2, the number of onboard passengers has already reached the maximum capacity, i.e., 1500 passengers. Similarly, train 2 and train 4 reach their maximum capacity at station 4 and station 3, respectively. So the trains are operated with high passenger

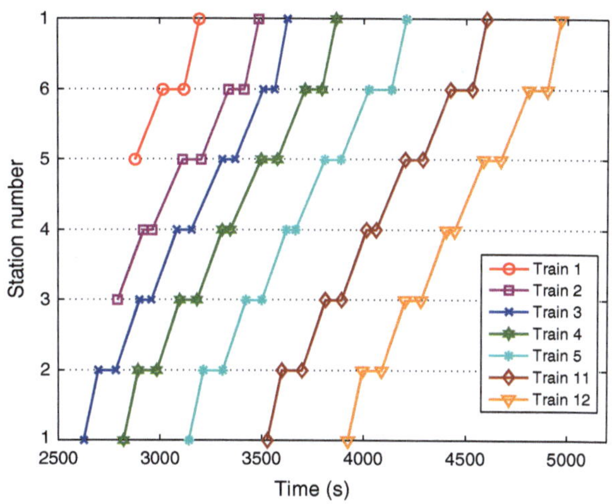

Fig. 7.10 Train schedules for line 1 obtained by the SQP method

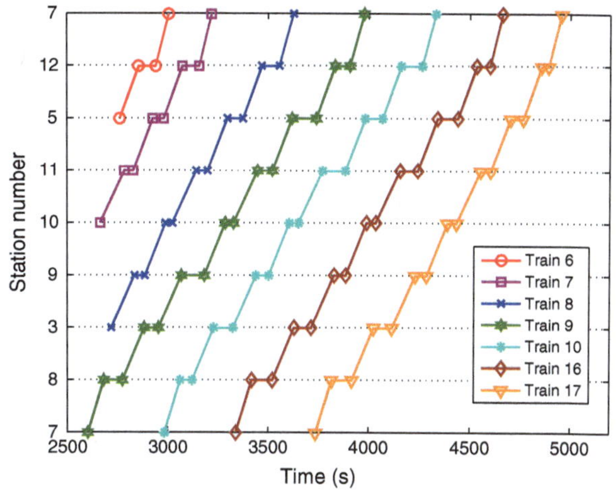

Fig. 7.11 Train schedules for line 2 obtained by the SQP method

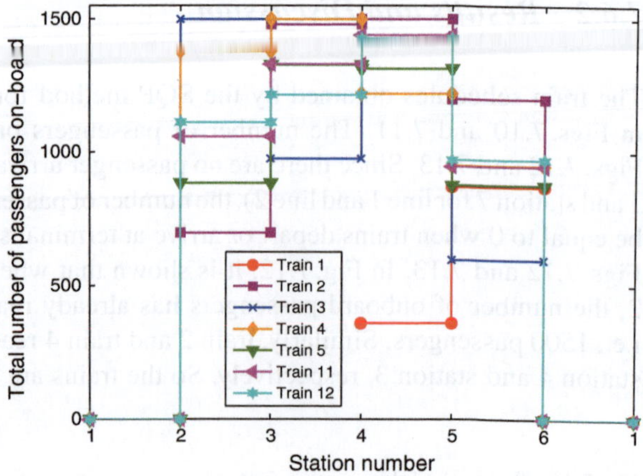

Fig. 7.12 Total number of onboard passengers obtained by the SQP method for line 1

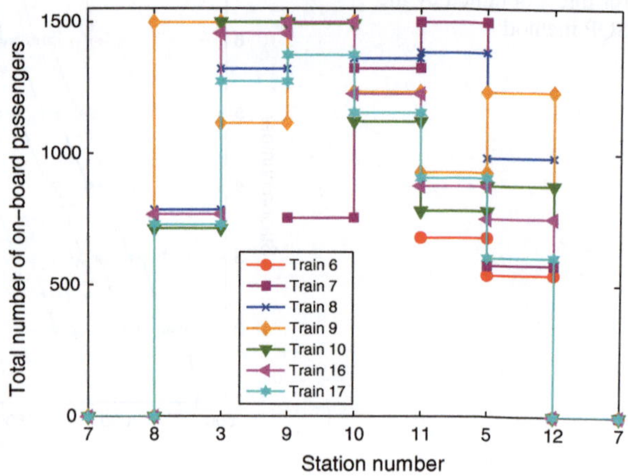

Fig. 7.13 Total number of onboard passengers obtained by the SQP method for line 2

load and the operation costs of these trains are smaller when compared with the operation of trains with lower passenger load.

As mentioned before, station 3 and station 5 are transfer stations in the small rail network shown in Fig. 7.9. Some passengers need to transfer to arrive at their destinations, e.g., passengers that enter the network at station 2 but have destination 9, 10, or 11 need to transfer at station 3. When trains of line 1 arriving at the platform of transfer station 3, the number of onboard passengers with different destinations is shown in Fig. 7.14, where the number of onboard passengers with destination 1, 2, 7, 8 is equal to 0. The passengers with destination 3 (dark purple bars) will get off the train at this station and the passengers with destination 9, 10, and 11 (pink bars) will also get off the train and transfer to line 2. The other passengers will stay on board. It is noted that the passengers with destination 12 choose to stay on the train instead of transferring to line 2 at station 3. This is because these passengers can also transfer at station 5 as shown in Fig. 7.15 and this will lead to a shorter travel time. Furthermore,

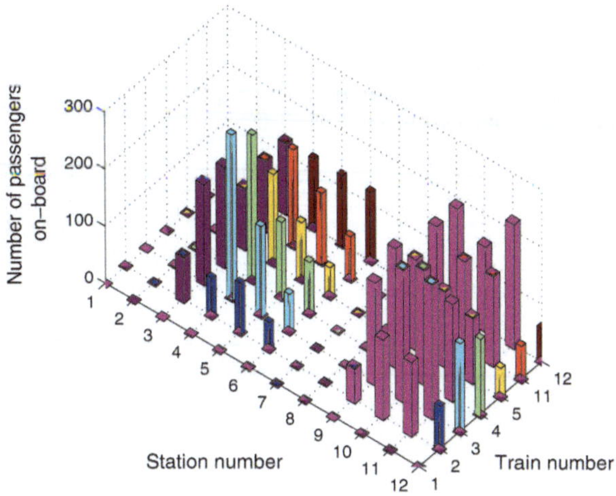

Fig. 7.14 Number of onboard passengers with different destinations at transfer station 3 obtained by the SQP method for line 1

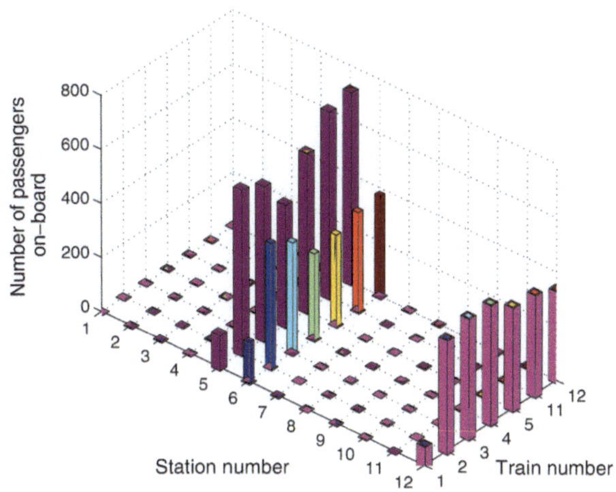

Fig. 7.15 Number of onboard passengers with different destinations at transfer station 5 obtained by the SQP method for line 1

the number of passengers at station 3 for train 1 is zero in Fig. 7.14, since train 1 has already passed station 3 at time t_0. Similarly, the number of onboard passengers with different destinations at station 3 and 5 of line 2 is shown in Figs. 7.16 and 7.17.

A comparison of the performance of the two approaches, i.e., the SQP method and the genetic algorithm, is illustrated in Table 7.6, where the values of the objective function, the computation time, the energy consumption of trains, the number and the travel time of the passengers that finished their trips, and the number and the waiting time of the passengers that did not travel are listed. It is observed that the SQP method has a better performance than the genetic algorithm for this case study. In particular, the objective value of the SQP method is about 10 % smaller than that obtained by the genetic algorithm. The train schedule obtained by the SQP method has a lower energy consumption and more passengers arrive at their destination. However, the

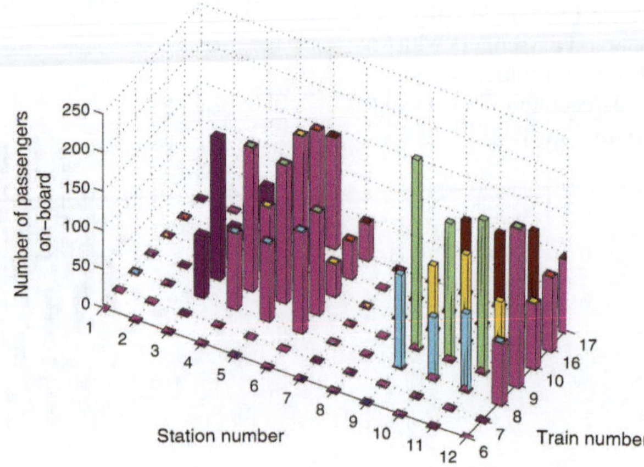

Fig. 7.16 Number of onboard passengers with different destinations at transfer station 3 obtained by the SQP method for line 2

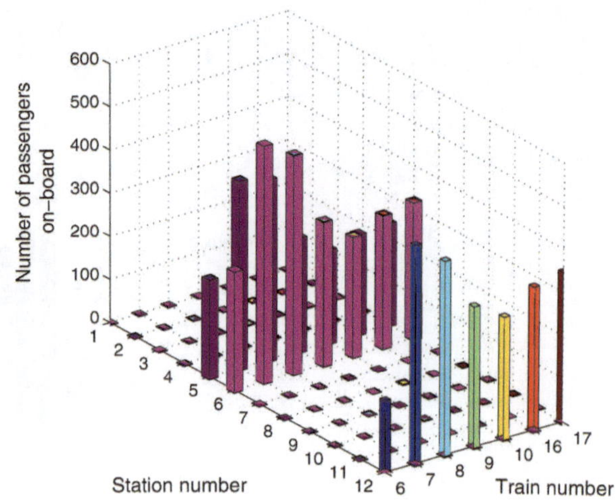

Fig. 7.17 Number of onboard passengers with different destinations at transfer station 5 obtained by the SQP method for line 2

travel time for passengers that finished their trip obtained by the genetic algorithm is smaller than that obtained by the SQP method. The reason for this is that the number of passengers that finished their trips in the schedule obtained by genetic algorithm is smaller than that of the SQP method. In addition, the higher energy consumption of the schedule obtained by genetic means less running time, so the travel time is also less.

For the given case study, the performance of the SQP method is about 10 % better that of the genetic algorithm. In addition, the SQP method also has a smaller computation time. In particular, for the small network with 2 lines given in Fig. 7.9, the computation time of the SQP method is about one hour using MATLAB on a 64-bit linux operation system running on 1.8 GHz Intel Core2 Duo CPU. Multiple experiments with different scenarios are need to access the performance of the solution approaches.

Table 7.6 Performance comparison of the SQP method and the genetic algorithm

Solution approaches	SQP method	Genctic algorithm
Objective value (–)	3.522	3.858
Computation time (s)	4.232×10^3	5.941×10^3
Energy consumption (J)	2.677×10^9	3.220×10^9
Number of passengers that finished their trips (passengers)	2.993×10^4	2.550×10^4
Number of passengers that did not finish their trips (passengers)	9.215×10^3	1.364×10^4
Travel time for passengers that finished their trips (s)	1.761×10^7	1.484×10^7
Waiting time for passengers that did not travel (s)	3.218×10^6	7.265×10^6

7.7 Summary

In this chapter, the train scheduling problem for an urban rail transit network is investigated. We have built an event-driven model with three types of events, i.e., departure events, arrival events, and passenger arrival rate change events at platforms. The splitting of passenger flows and passenger transfers at transfer stations are included in the event-driven model. For the given case study, the SQP method provides a better trade-off between control performance and computational complexity than the genetic algorithm.

An extensive comparison and assessment of the SQP method, the genetic algorithm, and other solution approaches for different scenarios will be a topic for future work. For the cases with multiple lines and a large number of stations and trains, distributed optimization approaches are expected to be applied to solve the train scheduling problem. In addition, we will also investigate the simplification of the proposed model so that the train scheduling problem can be solved in real time.

References

1. Wang Y, Tang T, Ning B, van den Boom T, De Schutter B (2015) Passenger-demands-oriented train scheduling for an urban rail transit network. Transp Res Part C 60:1–23
2. Wong S, Tong C (1998) Estimation of time-dependent origindestination matrices for transit networks. Transp Res Part B 32:35–48
3. Leemis L (1995) Reliability: probabilistic models and statistical methods. Prentice Hall, Englewood Cliffs
4. Bocharnikov Y, Tobias A, Roberts C, Hillmansen S, Goodman C (2007) Optimal driving strategy for traction energy saving on DC suburban railways. IEE Proc: Electr Power Appl 1:675–682

5. Ding Y, Bai Y, Liu F, Mao B (2009) Simulation algorithm for energy-efficient train control under moving block system. In: Proceedings of the 2009 World congress on computer science and information engineering, Los Angeles, CA, USA, pp 498–502
6. Yang X, Li X, Gao Z, Wang H, Tang T (2013) A cooperative scheduling model for timetable optimization in subway systems. IEEE Transp Intell Transp Syst 14:438–447

Chapter 8
Overview and Future Directions

Abstract In this book, we have discussed the optimal trajectory planning problem for trains and the real-time train scheduling problem for urban rail transit systems. The optimal trajectory planning problem for a single train and for multiple trains with constraints caused by signaling system has been investigated. Furthermore, the train scheduling based on passenger demand for urban rail transit systems has been considered in this book. There are still many issues to be tackled. However, the conflict between modeling accuracy and computation efficiency is one of the most important aspects that should be taken into account during the research.

In this book, we have discussed the optimal trajectory planning problem for trains and the real-time train scheduling problem for urban rail transit systems. The optimal trajectory planning problem for a single train and for multiple trains with constraints caused by signaling system has been investigated. Furthermore, the train scheduling based on passenger demand for urban rail transit systems has been considered in this book. There are still many issues to be tackled. However, the conflict between modeling accuracy and computation efficiency is one of the most important aspects that should be taken into account during the research. If the accuracy is increased, then the computational complexity will also increase, and vise versa. Hence, it is important to achieve a balanced trade-off between accuracy and computational efficiency. Furthermore, with the increasing complexity of the problem formulation, the computational complexity will also increase and it is difficult to solve the problem in real time using the current approaches. The modeling accuracy can be achieved by considering microscopic modeling, hybrid control methods for train trajectory planning, short-turning of trains, etc. Moreover, hierarchical optimization techniques, distributed optimization techniques, and fast optimization approaches could be adopted to increase the computational efficiency. In addition, urban rail transit networks are becoming more and more complex, especially in large cities like Beijing, Shanghai, Tokyo, and Paris. Passengers can arrive at their destinations through multiple available routes. Nowadays, rail operators can only give general route advices and passengers have the freedom to choose routes themselves. The passenger behavior, e.g., how passengers adapt to personalized route advices and to updated train schedules, can be investigated.

Y. Wang et al., *Optimal Trajectory Planning and Train Scheduling for Urban Rail Transit Systems*, Advances in Industrial Control,
DOI 10.1007/978-3-319-30889-0_8

The optimization models and algorithms proposed in this book can also be adapted to control and to optimize other transportation systems such as

- Bus transit systems.
  Bus transit systems with buses running on conventional roads to carry numerous passengers on short journeys, share many similarities with urban rail transit systems, such as variety in origin–destination passenger demands, uncertainty in dwell times and running times, and fixed routes. However, bus transit systems are operated with lower capacities, lower passenger demands, and more degrees of freedoms on the operations (e.g., overtaking, crossing, and turnaround). The model and solution approaches for train scheduling and trajectory planning provided in this book can be extended to bus transit systems, where the effect of other road traffic, e.g., cars and of traffic signals should be taken into account.
- Multi-car elevator systems.
  Elevator systems are usually controlled by a centralized upper-level controller, which determines where a car should stop to load or to unload passengers. Similar as the train scheduling considering passenger demands, elevator dispatching is also characterized by time-varying passenger arrival patterns. Therefore, the scheduling and trajectory planning methods presented in this book can be applied to reduce the waiting time of passengers and the energy consumption.
- Automated guided vehicles (AGVs).
  AGVs are automated vehicles that can load, unload, and transport goods in port container terminals or manufacturing systems. In general, there are two types of AGVs: one type follows prescribed routes (indicated by markers or wires in the floor) or it has to select its route from a prescribed route set, while the other type uses vision or lasers for navigation and it can go to anywhere in the specified area. In addition, energy consumption is very crucial for AGVs with battery or fuels since they could run out of energy. The models and solution algorithms presented in this book can also be extended for AGVs with prescribed routes, in particular the limited-energy aspect should be included in the scheduling problem.

Appendix A
A General Formulation of the Pseudospectral Method

The optimal train trajectory planning problem of Chap. 3 can be formulated as a multiple phase optimal control problem and then can be solved by the pseudospectral method. Below we describe a general optimal control problem with multiple phases and the solution procedure of the general optimal control problem using the pseudospectral method. This explanation is based on [1–5].

A.1 The Multiple-Phase Optimal Control Problem

The general optimal control problem with N_p phases is formulated as follows [2].

The objective function is

$$J = \sum_{i=1}^{N_\text{p}} \left(\varphi^{(i)}\big(x^{(i)}(t_\text{f}^{(i)}), p^{(i)}, t_\text{f}^{(i)}\big) + \int_{t_0^{(i)}}^{t_\text{f}^{(i)}} L^{(i)}\big(x^{(i)}(t), u^{(i)}(t), p^{(i)}, t\big)\text{d}t \right), \quad \text{(A.1)}$$

where $[t_0^{(i)}, t_\text{f}^{(i)}]$ is the time interval for the ith phase, $u^{(i)}(\cdot)$ and $x^{(i)}(\cdot)$ are the control trajectories and state trajectories, $p^{(i)}$ are the static parameters, for $i = 1, 2, \dots, N_\text{p}$.

The objective function (A.1) is subject to the differential constraints

$$\dot{x}^{(i)}(t) = f^{(i)}\big(x^{(i)}(t), u^{(i)}(t), p^{(i)}, t\big), \quad t \in [t_0^{(i)}, t_\text{f}^{(i)}], \quad \text{(A.2)}$$

the path constraints

$$h_\text{L}^{(i)} \le h^{(i)}\big(x^{(i)}(t), u^{(i)}(t), p^{(i)}, t\big) \le h_\text{U}^{(i)}, \quad t \in [t_0^{(i)}, t_\text{f}^{(i)}], \quad \text{(A.3)}$$

Y. Wang et al., *Optimal Trajectory Planning and Train Scheduling for Urban Rail Transit Systems*, Advances in Industrial Control,
DOI 10.1007/978-3-319-30889-0

the event constraints

$$
\begin{aligned}
e_{\mathrm{L}}^{(i)} \le e^{(i)}\big(x^{(i)}(t_0^{(i)}), u^{(i)}(t_0^{(i)}), x^{(i)}(t_{\mathrm{f}}^{(i)}), u^{(i)}(t_{\mathrm{f}}^{(i)}), p^{(i)}, t_0^{(i)}, t_{\mathrm{f}}^{(i)}\big) \le e_{\mathrm{U}}^{(i)},\\
t \in [t_0^{(i)}, t_{\mathrm{f}}^{(i)}],
\end{aligned} \tag{A.4}
$$

the linkage constraints

$$
\begin{aligned}
\Psi_{\mathrm{L}} \le \Psi\big(&x^{(1)}(t_0^{(1)}), u^{(1)}(t_0^{(1)}), x^{(1)}(t_{\mathrm{f}}^{(1)}), u^{(1)}(t_{\mathrm{f}}^{(1)}), p^{(1)}, t_0^{(1)}, t_{\mathrm{f}}^{(1)},\\
&x^{(2)}(t_0^{(2)}), u^{(2)}(t_0^{(2)}), x^{(2)}(t_{\mathrm{f}}^{(2)}), u^{(2)}(t_{\mathrm{f}}^{(2)}), p^{(2)}, t_0^{(2)}, t_{\mathrm{f}}^{(2)},\\
&\ldots\\
&x^{(N_{\mathrm{p}})}(t_0^{(N_{\mathrm{p}})}), u^{(N_{\mathrm{p}})}(t_0^{(N_{\mathrm{p}})}), x^{(N_{\mathrm{p}})}(t_{\mathrm{f}}^{(N_{\mathrm{p}})}), u^{(N_{\mathrm{p}})}(t_{\mathrm{f}}^{(N_{\mathrm{p}})}), p^{(N_{\mathrm{p}})}, t_0^{(N_{\mathrm{p}})}, t_{\mathrm{f}}^{(N_{\mathrm{p}})}\big) \le \Psi_{\mathrm{U}},
\end{aligned} \tag{A.5}
$$

the bound constraints

$$
\begin{aligned}
u_{\mathrm{L}}^{(i)} \le u^{(i)}(t) \le u_{\mathrm{U}}^{(i)}, \quad x_{\mathrm{L}}^{(i)} \le x^{(i)}(t) \le x_{\mathrm{U}}^{(i)}, \quad t \in [t_0^{(i)}, t_{\mathrm{f}}^{(i)}],\\
p_{\mathrm{L}}^{(i)} \le p^{(i)} \le p_{\mathrm{U}}^{(i)}, \quad \underline{t}_0^{(i)} \le t_0^{(i)} \le \bar{t}_0^{(i)}, \quad \underline{t}_{\mathrm{f}}^{(i)} \le t_{\mathrm{f}}^{(i)} \le \bar{t}_{\mathrm{f}}^{(i)},
\end{aligned} \tag{A.6}
$$

and the time constraints

$$
t_{\mathrm{f}}^{(i)} - t_0^{(i)} \ge 0. \tag{A.7}
$$

A.2 The Solution Process of the Optimal Control Problem

Let $i \in \{1, 2, \ldots, N_{\mathrm{p}}\}$ be a particular phase of the optimal control problem (A.1)–(A.7) and let $(\cdot)^{(i)}$ denote information for the ith phase. The ith phase of the optimal control problem can be transformed from the interval $t \in [t_0^{(i)}, t_{\mathrm{f}}^{(i)}]$ to the interval $\tau \in [-1, 1]$ for $i = 1, 2, \ldots, N_{\mathrm{p}}$ by introducing the following transformation [2]:

$$
\tau = \frac{2}{t_{\mathrm{f}}^{(i)} - t_0^{(i)}} t - \frac{t_{\mathrm{f}}^{(i)} + t_0^{(i)}}{t_{\mathrm{f}}^{(i)} - t_0^{(i)}}. \tag{A.8}
$$

Now we approximate the state and control functions using Legendre pseudospectral approximation. The state $x_k^{(i)}(\tau)$, $\tau \in [-1, 1]$ is approximated by the N_ith order Lagrange polynomial $x_k^{N_i,(i)}(\tau)$ based on interpolation at the Legendre-Gauss-Lobatto (LGL) points [4]:

$$
x_k^{(i)}(\tau) \approx x_k^{N_i,(i)}(\tau) = \sum_{n=0}^{N_i} \bar{x}_k^{N_i,(i)}(\tau_n)\phi_n^{(i)}(\tau), \tag{A.9}
$$

where $\bar{x}_k^{N_i,(i)}(\tau_n)$ is a discrete approximation of the LGL point τ_n and the Lagrange basis polynomials $\phi_n^{(i)}(\tau)$ for $n = 0, 1, \cdots, N_i$ are defined as

$$\phi_n^{(i)}(\tau) = \prod_{m=0, m\neq n}^{N_i} \frac{\tau - \tau_m^{(i)}}{\tau_n^{(i)} - \tau_m^{(i)}}, \tag{A.10}$$

and $\tau_n^{(i)}$ for $n = 0, 1, \ldots, N_i$ are the LGL points, which are defined as $\tau_0^{(i)} = -1$, $\tau_{N_i}^{(i)} = 1$, and τ_n for $n = 1, 2, \ldots, N_i - 1$ being the subsequent roots of the derivative of the Legendre polynomial

$$L_{N_i}(\tau) = \frac{1}{2^{N_i} N_i!} \frac{\mathrm{d}^{N_i}}{\mathrm{d}\tau^{N_i}} (\tau^2 - 1)^{N_i},$$

in the interval $[-1, 1]$. The control $u^{(i)}(\tau)$ can be approximated in a similar way. The derivative of $x_k^{N_i,(i)}(\tau)$ at the LGL points τ_n can be obtained by differentiating (A.9), which can be expressed as a matrix multiplication as follows:

$$\dot{x}_k^{(i)}(\tau_n) \approx \dot{x}_k^{N_i,(i)}(\tau_n) = \sum_{j=0}^{N_i} \tilde{D}_{nj}^{(i)} \bar{x}_k^{N_i,(i)}(\tau_j), \tag{A.11}$$

where $\tilde{D}^{(i)}$ is the $(N_i + 1) \times (N_i + 1)$ differential approximation matrix [6] given by

$$\tilde{D}_{nj}^{(i)} = \begin{cases} \frac{\phi_{N_i}^{(i)}(\tau_n)}{\phi_{N_i}^{(i)}(\tau_j)} \frac{1}{\tau_n - \tau_j}, & \text{if } n \neq j, \\ -N_i(N_i + 1)/4, & \text{if } n = j = 0, \\ N_i(N_i + 1)/4, & \text{if } n = j = N_i, \\ 0, & \text{otherwise.} \end{cases} \tag{A.12}$$

The differential constraints can be recast into algebraic constraints via the differential approximation matrix. In addition, the path constraints (A.3) can be discretized at the LGL points. Note that the dynamic constraints and path constraints are only considered at the LGL points, which means both the dynamic and path constraints might be violated in between the LGL points [7]. The objective function (A.1) can be approximated using the LGL points as

$$J = \sum_{i=1}^{N_p} \Bigg(\varphi^{(i)}\left(\bar{x}^{N_i,(i)}(-1), \bar{x}^{N_i,(i)}(1), p^{(i)}, t_0^{(i)}, t_f^{(i)}\right) + \frac{t_f^{(i)} - t_0^{(i)}}{2} \sum_{n=0}^{N_i} L^{(i)}\left(\bar{x}^{N_i,(i)}(\tau_n), \bar{u}^{N_i,(i)}(\tau_n), p^{(i)}, \tau_n\right)\omega_n \Bigg), \tag{A.13}$$

where ω_n are weights given by

$$\omega_n = \frac{2}{N(N+1)} \frac{1}{(L_{N_i}(\tau_n))^2}, \quad \text{for } n = 0, 1, \ldots, N_i. \tag{A.14}$$

If we include all the decision variables in vector y, the optimal control problem can then be expressed as a nonlinear programming problem:

$$\min_y J(y), \tag{A.15}$$

subject to

$$G_\mathrm{L} \le G(y) \le G_\mathrm{U}$$
$$y_\mathrm{L} \le y \le y_\mathrm{U}. \tag{A.16}$$

By defining $G(y)$, G_L, G_U, y_L, and y_U properly, we can write all constraints in the form (A.16).

There exist several commercial and free packages that implement the pseudospectral method: SOCS [8] and DIRCOL [9] are Fortran-based proprietary packages, while PROPT [7] and DIDO [10] are commercial software packages with MATLAB® interface. A MATLAB-based open source tool that uses the Gauss pseudospectral method is GPOPS [11]. PSOPT is an open source optimal control package written in C++, including Legendre and Chebyshev pseudospectral discretizations [1]. These software packages start with a general optimal problem formulated as (A.1)–(A.7), then transform this problem into a nonlinear programing problem with objective function (A.15) and constraints (A.16), and finally solve it possibly using an NLP solver, such as SNOPT [12].

Appendix B
Background—Optimization

In this appendix, background material on optimization (or mathematical programming) is presented. In Sect. B.1 a brief introduction of optimization is presented. Solution approaches for nonlinear programming problems are given in Sect. B.2. In addition, solution approaches for mixed-integer programming problems are introduced shortly in Sect. B.3.

B.1 Introduction of Optimization

Many real-world optimization problems from mechanical engineering, control engineering, economics, etc. can be modeled using the following general form [13]:

$$\begin{aligned} &\min_x f(x) \\ &\text{s.t. } h(x) = 0 \text{ and } g(x) \leq 0, \\ &x = \left[x_{\text{in}}^T \; x_{\text{re}}^T\right]^T, \; x_{\text{in}} \in \mathbb{Z}^{n_{\text{in}}}, \; x_{\text{re}} \in \mathbb{R}^{n_{\text{re}}}. \end{aligned} \tag{B.1}$$

where $\mathbb{Z}$ and $\mathbb{R}$ are the sets of integers and real numbers respectively, $f(\cdot)$ is the objective function of the optimization problem, x is the parameter vector that can be used to minimize the objective function, $h(\cdot) = 0$ and $g(\cdot) \leq 0$ are the equality and inequality constraints that restrict the solution to a certain subset of the parameter space. Note that there is no fundamental difference between maximization and minimization; so we only consider minimization here. For an optimization problem, the objective function $f(\cdot)$ and the parameter vector x are always present. However, the elements $g(\cdot)$ and $h(\cdot)$ can be omitted if no constraints are imposed on the values of the parameters. The unconstrained optimization problem and constrained problem can be distinguished based on the presence or absence of the constraints. For a

Y. Wang et al., *Optimal Trajectory Planning and Train Scheduling for Urban Rail Transit Systems*, Advances in Industrial Control,
DOI 10.1007/978-3-319-30889-0

constrained problem, if a point x satisfy the constraints, i.e., $h(x) = 0$ and $g(x) \leq 0$, then the point x is called feasible; otherwise, the point x is called infeasible.

Furthermore, the optimization problems can also be distinguished based on the parameter vector x, objective function $f(\cdot)$, and constraints $h(\cdot)$ and $g(\cdot)$. As given in (B.1), the parameters in x could be integer variables and/or real-valued variables. If the optimization problem is a real-valued optimization problem, then x_{in} in (B.1) does not exist. For a real-valued optimization problem, we have the following three cases:

- Linear programming problem: $f(\cdot)$ is a linear function and the constraints $g(\cdot)$ and $h(\cdot)$ are also linear[1] functions.
- Quadratic programming problem: $f(\cdot)$ is a quadratic function and the constraints $g(\cdot)$ and $h(\cdot)$ are linear functions.
- Nonlinear programming problem: $f(\cdot)$ is a nonlinear function (non quadratic if the constraints are linear) and/or the constraints $g(\cdot)$ and/or $h(\cdot)$ are nonlinear functions.

If x has integer values only, i.e., x_{re} does not exist, then the optimization problem is a pure integer optimization problem. Furthermore, if both x_{re} and x_{in} are present, then the optimization problem is a mixed-integer optimization problem.

If both the objective function and the feasible set are convex, then the minimum obtained is a global minimum; otherwise, there may be several local minima. A local minimum x^* is defined as a point for which exists some $\delta > 0$ so that for all x such that

$$\|x - x^*\| \leq \delta, \tag{B.2}$$

the expression

$$f(x^*) \leq f(x), \tag{B.3}$$

holds, i.e., on a certain region around x^* all the function values are larger than or equal to the function value of x^* [13].

The optimization problems considered in this book are mostly nonlinear programming and mixed integer nonlinear programming problems. We solve these optimization problems via sequential quadratic programming, iterative convex programming, genetic algorithms, pattern search, branch-and-bound, multi-start method, and bi-level optimization method. A brief introduction for these methods used is given next.

---

[1] A function f is affine if it is of the form $f(x) = a_1x_1 + a_2x_2 + \cdots + a_nx_n + b = a^Tx + b$ with $a \in \mathbb{R}^n$ and $b \in \mathbb{R}$. If $b = 0$ then we say that the function is linear. Note that many people also use the phrases "linear" even if $b \neq 0$. Here we also use "linear" in this appendix.

B.2 Solution Approaches for Nonlinear Programming Problems

B.2.1 Sequential Quadratic Programming

Sequential quadratic programming (SQP) is an iterative method for real-valued nonlinear programming problems that has been widely used to solve different nonlinear programming problems in practice [14]. The SQP method belongs the category of gradient-based nonlinear programming methods, which rely on gradient or Hessian information. If this information is not available, it can be approximated numerically. So a requirement for the SQP algorithm is that the objective function and the constraints of the nonlinear programming problem should be continuously differentiable [14]. In the SQP method, the nonlinear programming problem is recast as a sequence of quadratic programming problems, which can be solved easily and efficiently by an interior point method, an active set method, etc.

For the sake of simplicity of the exposition, we consider a nonlinear optimization problem in (B.1) but only with equality constraints [13, 15]. A Lagrange function is introduced as follows:

$$L(x, \lambda, \sigma) = f(x) + \lambda^T h(x), \tag{B.4}$$

where the parameter vector x is in $\mathbb{R}^n$, $\lambda \in \mathbb{R}^p$ are the Lagrange multipliers for the equality constraints. The Kuhn-Tucker necessary conditions for optimality can be stated as

$$F(x, \lambda) = \begin{bmatrix} \nabla f(x) + \nabla h(x)\lambda \\ h(x) \end{bmatrix} = 0, \tag{B.5}$$

where $\nabla h(x)$ is the Jacobian matrix of the equality constraints $h(x)$ and is an $n \times p$ matrix. So $F(x, \lambda)$ is a set of $n+p$ nonlinear equations with $n+p$ unknown variables, i.e., $x \in \mathbb{R}^n$ and $\lambda \in \mathbb{R}^p$. These nonlinear equations can be solved using Newton's method. The Jacobian matrix of $F(x, \lambda)$ can be written as

$$\nabla_{x,\lambda} F(x, \lambda) = \begin{bmatrix} \nabla^2_{x,x} L(x, \lambda) & \nabla h(x) \\ \nabla^T h(x) & 0 \end{bmatrix}. \tag{B.6}$$

According to the Newton's method, the solution of (B.5) can be found iteratively as

$$\begin{bmatrix} x_{k+1} \\ \lambda_{k+1} \end{bmatrix} = \begin{bmatrix} x_k \\ \lambda_k \end{bmatrix} + \begin{bmatrix} \Delta x_k \\ \Delta \lambda_k \end{bmatrix}, \tag{B.7}$$

where Δx_k and $\Delta \lambda_k$ are obtained by solving

$$\begin{bmatrix} \nabla^2_{x,x} L(x_k, \lambda_k) & \nabla h(x_k) \\ \nabla^T h(x_k) & 0 \end{bmatrix} \begin{bmatrix} \Delta x_k \\ \Delta \lambda_k \end{bmatrix} = \begin{bmatrix} -\nabla f(x_k) - \nabla h(x_k)\lambda_k \\ -h(x_k) \end{bmatrix}. \tag{B.8}$$

By substituting $\Delta\lambda_k = \lambda_{k+1} - \lambda_k$ given in (B.7) into (B.8), we have

$$\begin{bmatrix} \nabla^2_{xx} L(x_k, \lambda_k) & \nabla h(x_k) \\ \nabla^T h(x_k) & 0 \end{bmatrix} \begin{bmatrix} \Delta x_k \\ \lambda_{k+1} \end{bmatrix} = \begin{bmatrix} -\nabla f(x_k) \\ -h(x_k) \end{bmatrix}. \tag{B.9}$$

The iterative process indicated by (B.7) and (B.9) can be continued until the convergence is achieved.

We consider the following quadratic programming problem:

$$\min_{\Delta x_k} = \nabla^T f(x_k)\Delta x_k + \frac{1}{2}\Delta x_k^T \nabla^2_{x,x} L(x_k, \lambda_k)\Delta x_k$$
$$\text{s.t. } h(x_k) + \nabla^T h(x_k)\Delta x = 0. \tag{B.10}$$

The Lagrange function of the quadratic programming problem can be defined as

$$\tilde{L} = \nabla^T f(x_k)\Delta x_k + \frac{1}{2}\Delta x_k^T \nabla^2_{x,x} L(x_k, \lambda_k)\Delta x_k + \tilde{\lambda}_k^T (h(x_k) + \nabla^T h(x_k)), \tag{B.11}$$

where $\tilde{\lambda}_k$ is the Lagrange multiplier. The Kuhn-Tucker necessary conditions can be written as

$$\nabla f(x_k) + \nabla^2_{x,x} L(x_k, \lambda_k)\Delta x_k + \nabla h(x_k)\tilde{\lambda}_k = 0, \tag{B.12}$$
$$h(x_k) + \nabla^T h(x_k)\Delta x_k = 0. \tag{B.13}$$

It is noted that (B.12) and (B.13) can be identified to be the same as (B.9). So the original nonlinear problem of (B.1) but only with equality constraints can be solved in an iterative way by solving the quadratic programming problem defined by (B.10).

The SQP procedure for the equality-constrained nonlinear programing can be extended to a general nonlinear programming problem with both equality and inequality constraints [13, 15]. Note that the Hessian of the Lagrangian $\nabla^2_{x,x} L(x_k, \lambda_k)$ is made up of the second derivatives of the objective function and constraints, which may be time-consuming. So it is useful to consider replacing the Hessian $\nabla^2_{x,x} L(x_k, \lambda_k)$ by a quasi-newton approximation, e.g., the Broyden-Fletcher-Goldfarb -Schanno approximation [15]. SQP methods have been implemented in many commercial and free software packages, such as SNOPT [12, 16], NPSOL [17], and MATLAB [18].

B.2.2 Pattern Search

Pattern search is a direct-search method [19], which does not require the gradient or Hessian of the objective function and of the constraints to solve the optimization problem. Therefore, pattern search can be applied to functions that are not continuous nor differentiable. Pattern search methods [15] choose a certain set of search directions at each iteration point and evaluate the objective function at a given step length along each of these directions. If a point with a significantly lower function value is found, then this point is adopted as the new iteration point, where the set of directions may be changed and the step length may grow or shrink as explained next.

We define x_k to be the solution at iterative step k, D_k to be the set of possible search directions, and ℓ_k to be the step size. Then the candidates for the current iteration are $x_k + \ell_k d_k$ for all $d_k \in D_k$. In addition, a function $\rho(\cdot)$ of the step length is introduced to evaluate the decrease of the objective function. The iterative procedure of pattern search is given as follows [15]:

- Choose initial point x_0, initial step length ℓ_0, and initial direction set D_0. Note that the initial step length ℓ_0 should be larger than the convergence tolerance step length ℓ_{tol}.
- Evaluate the candidates $x_k + \ell_k d_k$ for all $d_k \in D_k$ for each iteration k.
  - If $f(x_k + \ell_k d_k) \leq f(x_k) - \rho(\ell_k)$ for some $d_k \in D_k$, then set $x_{k+1} = x_k + \ell_k d_k$ for such d_k. In addition, also set $\ell_{k+1} = \phi_k \ell_k$ with $\phi_k \geq 1$ to increase the step length.
  - If the inequality $f(x_k + \ell_k d_k) \leq f(x_k) - \rho(\ell_k)$ cannot be satisfied for any of the candidates, then we set $x_{k+1} = x_k$ and reduce the step length by $\ell_{k+1} = \theta_k \ell_k$ with $0 < \theta_k < 1$.
- Terminate the iterative procedure if the step length ℓ_k is less than the given convergence tolerance ℓ_{tol}.

Note that the pattern search iterative procedure does not require us to choose the point $x_k + \ell_k d_k$ with the smallest objective value, so the function of evaluations could be reduced by not evaluating the function values for all the candidates, but performing the evaluations one at a time and accepting the first candidate point that satisfies the sufficient decrease condition.

Several pattern search algorithms have been implemented in the Global Optimization Toolbox of the MATLAB [20], such as the generalized pattern search algorithm, the generating set search algorithm, and the mesh adaptive search algorithms.

B.2.3 Genetic Algorithm

Genetic algorithms are based on the concepts of biological evolution, which involves the principles of natural genetics and nature selection [21, 22]. The basic elements of nature genetics are reproduction, crossover, and mutation, which are used in genetic search procedure. So genetic algorithms do not depend on the gradient and/or Hessian information, but only based on the function evaluations. Many practical optimization problems, such as non-convex nonlinear programming problems, mixed integer nonlinear programming problems, can be solved by genetic algorithms and in some cases the global optimum solution can be obtained with a high probability [13].

In genetic algorithms, a set of possible candidate solutions, i.e. a population, is used for starting the procedure, where each candidate is represented by a string of binary variables corresponding to the chromosomes in nature genetics. For continuous variables, the desired resolution can be achieved by varying the string length. Since a whole population of possible solutions is considered, genetic algorithms can escape from local minima. Each candidate in the initial population is evaluated to obtain its fitness value. Then three genetic operators, i.e., selection, crossover, and mutation, are operated to produce a new population of solutions. The new population is then evaluated to get the fitness values and to obtain the convergence of the iterative process. If the convergence criterion or the maximum number of generations is reached, then the iterative procedure terminates; otherwise, the population is iteratively operated by the three operators and the resulting new population is evaluated.

By adopting the survival-of-the-fittest principle, genetic algorithms try to maximize the fitness function $F(\cdot)$. The fitness function can be constructed by the objective function $f(\cdot)$, the equality constraints $h(x) = 0$, and the inequality constraints $g(x) \leq 0$ of the minimization problem given in (B.1). A possible construction of the fitness function $F(\cdot)$ could be [13]

$$F(x) = \frac{1}{1 + \phi(x)}, \tag{B.14}$$

where

$$\phi(x) = f(x) + \alpha h(x) h^T(x) + \beta \hat{g}(x) \hat{g}^T(x), \tag{B.15}$$

with

$$\hat{g}_i(x) = \begin{cases} g_i(x) & \text{if } g_i(x) > 0, \\ 0 & \text{if } g_i(x) \leq 0, \end{cases} \quad \text{for } i = 1, 2, \ldots, m \tag{B.16}$$

and where α and β are the penalty parameters associated with the constraints $h(x) = 0$ and $g(x) = 0$, and m is the total number of inequality constraints.

Furthermore, the three genetic operators are explained in detail as follows [21, 22]:

- Selection. The selection operator is used to pick the good strings from the current population and to put their multiple copies in the mating pool based on a probabilistic procedure. In the selection process, the strings with a higher fitness value will be selected more frequently and get more copies in the mating pool because of the larger probability and vice versa. Note that no new strings are produced in the selection phase, but only the existing candidates are copied in the mating pool.
- Crossover. After the selection process, the crossover operator is applied to generate new strings, where some parts of the strings are exchanged between two randomly picked individual strings in the mating pool. The two strings participating in the crossover operator are called parents strings and the strings created by crossover

process are child strings. The child strings generated via a random crossover that could be better or worse than their parents. If they are better than their parents, the average fitness of the new population will be improved faster. However, if they are worse than their parents, they would be disregarded in the next selection process.

- Mutation. The mutation operator is applied to the new population with a small mutation probability, where one or more binary variables of a string changes from 1 to 0 and vice versa. Through the mutation process, a string in the neighborhood of the current string is generated to accomplish a local search around the current solution and to maintain diversity of the population.

The computational procedure of genetic algorithm includes the following steps:

1. **Initialization**: Choose a proper string length and create an initial population. Set appropriate values for the crossover probability, mutation probability, etc.
2. **Iteration**: Evaluate the fitness values for the candidate solutions in the population and carry out the selection process, crossover process, and mutation process to create a new population.
3. **Stopping criterion**: The optimization process stops if the maximum number of generations is reached or the convergence criterion is satisfied.

B.2.4 Iterative Convex Programming

Inspired by the iterative quadratic programming approach proposed in [23], the iterative convex programming (ICP) approach is proposed to solve nonlinear non-convex optimization problems with real-valued variables approximately [24]. In the ICP approach, the variables x of the nonlinear non-convex problem given in (B.1) are split into two vectors x_1 and x_2, so the optimization problem (B.1) can be written as:

$$\begin{aligned} &\min_{x_1, x_2} f(x_1, x_2) \\ &\text{s.t. } h(x_1, x_2) = 0, \\ &\quad\; g(x_1, x_2) \le 0, \end{aligned} \tag{B.17}$$

where $x = \left[x_1^T \; x_2^T\right]^T$. The optimization problem in (B.17) can be split in to the following two subproblems:

$$\begin{aligned} &\min_{x_1} f(x_1, \hat{x}_2) \\ &\text{s.t. } h(x_1, \hat{x}_2) = 0, \\ &\quad\; g(x_1, \hat{x}_2) \le 0, \end{aligned} \tag{B.18}$$

and

$$
\begin{aligned}
&\min_{x_2} f(\hat{x}_1, x_2) \\
&\text{s.t. } h(\hat{x}_1, x_2) = 0, \\
&\qquad g(\hat{x}_1, x_2) \leq 0,
\end{aligned} \tag{B.19}
$$

where $\hat{x}_1$ and $\hat{x}_2$ are not variables any more, but are fixed values in the optimization models. If both (B.18) and (B.19) are convex programming problems, then the ICP approach can be applied to solve this nonlinear non-convex problem. A convex programming problem can be solved for the global optimum by use of efficient algorithms such as interior point methods [15]. However, the simplifications for the ICP approach mean that the global optimum of the convex problem is not necessarily the global minimum of the original nonlinear non-convex problem. Therefore, an iterative process is adopted to reduce the errors introduced by the simplifications. The iterative procedure can be described as the following steps:

1. Obtain the initial estimates $\hat{x}_2(0)$ of x_2 for the nonlinear non-convex problem.
2. Based on the estimates $\hat{x}_2(k)$, perform the convex programming for (B.18) to find the estimates $\hat{x}_1(k)$.
3. Based on the new estimate $\hat{x}_1(k)$ and perform the convex programming for (B.19) to find estimates $\hat{x}_2(k+1)$.
4. Repeat step 2 and 3 until the maximum number of iterations is reached or the convergence of optimization variables is achieved.

Remark B.1 When the programming problem is nonlinear and non-convex, there could exist multiple local minima. So the nonlinear programming approaches introduced before should be combined with the multi-start methods to obtain a solution close to a globally optimal solution. Multi-start methods generally increase the computation time required to solve an optimization problem significantly; however, multi-start methods can typically be executed in a highly parallel manner. In particular, when a straightforward multi-start method is chosen that relies on randomly generated initial solutions, then each optimization problem involved in the multi-start method can be solved on an independent processor.

B.3 Solution Approaches for Mixed-Integer Optimization Problem

Mixed-integer programming problems have both real-valued and integer-valued parameters as given in (B.1). Sections B.3.1 and B.3.2 introduce some solution approaches for mixed integer linear programming problems and mixed integer nonlinear programming problems respectively.

B.3.1 Mixed Integer Linear Programming

There are several advanced algorithms available for solving mixed integer linear programming (MILP) problems, such as the cutting plane algorithm [25] and the branch-and-bound method [26]. In particular, the branch-and-bound method has been applied widely for solving MILP problems because of its effectiveness. Hence, a detailed description of the branch-and-bound method is introduced next.

For the branch-and-bound method, a linear programming relaxation problem of an MILP problem is first solved without considering the integer constraints. If the values of the integer variables happen to be integers, then the MILP problem has been solved successfully: the optimal solution of the linear programming relaxation problem is the optimal solution of the original MILP problem. Otherwise, the optimal solution for the MILP problem can be obtained as follows. Let x_i be an integer variable having a real value in the solution of the linear programming relaxation problem. Then we can find an integer $\hat{x}_i$ that satisfies

$$\hat{x}_i \leq x_i \leq \hat{x}_i + 1. \tag{B.20}$$

With (B.20), the linear programming relaxation problem can be branched on the variable x_i and two subproblems are created as follows [13]:

1. The linear programming relaxation problem with constraint $x_i \leq \hat{x}_i$.
2. The linear programming relaxation problem with constraint $x_i \geq \hat{x}_i + 1$.

Note that the branching process can keep all the feasible integer solutions for MILP problems. Each of the two subproblems can then be solved as a real-valued linear programming problem and the solution of these two subproblems can be branched further on other integer variables. This branching process continues until all $x_{\text{in}} \in \mathbb{Z}^{n_{\text{in}}}$. When such a feasible solution is found, the corresponding objective function value is set as an upper bound for the objective function. Then the branching nodes of which the objective function values are larger than the upper bound can be cut or eliminated because further branching on these nodes will result in bigger objective function values. In addition, the upper bound of the objective function is updated when feasible solution with a smaller objective function value is found. Eventually, all branches except the optimal solution are eliminated from consideration and the branch-and-bound procedure is terminated.

The branch-and-bound procedure can be efficient for eliminating non-optimal candidate solutions because several branches could be cut (i.e., the subproblems on these branch should not be solved anymore) when a upper bound of the objective function is obtained. A branch can be cut if one of the following conditions is satisfied [13]:

- The subproblem does not have a feasible solution.
- The optimal solution of the subproblem is feasible for the original MILP problem.
- The objective function value of the subproblem is larger than the current upper bound.

The MILP problem can be solved by several existing commercial and free solvers [27–29], such as CPLEX, GUROBI, GLPK, LINDO, LINGO, SCIP.

B.3.2 Mixed Integer Nonlinear Programming

The extension of the branch-and-bound algorithm described in Sect. B.3.1 for solving mixed integer nonlinear programming (MINLP) problems is straightforward, where nonlinear programming subproblems have to be solved instead of linear programming subproblems [30]. If the relaxed real-valued nonlinear optimization problem is convex, the MINLP algorithm will converge to the global optimum [30].

The MINLP problems as well as nonlinear programming problems can be solved via a MILP approach, where the nonlinear terms can be approximated by piecewise affine functions. Then the original nonlinear programming problem is transformed into an MILP problem by applying transformation properties [31]. It is noted that the number of binary variables introduced by the nonlinear terms is related to the number of piecewise affine subfunctions. So attention should be paid to keeping the number of binary variables small when reformulating the MINLP and nonlinear programming problems into MILP problems; because the MILP problem has been proven to be NP-hard and increasing the number of variables will increase the computational complexity. However, fewer binary variables would lead to fewer affine pieces in the approximation process, which may cause larger discrepancy with the original nonlinear functions. Therefore, it is important to find a good trade-off between accuracy of approximation and computational complexity. Moreover, the optimal solution of the MILP problems may differ from that of the MINLP problems because of the error caused by the approximations.

In addition, a bi-level optimization method consisting of two levels of optimization could be used for solving the MINLP problems, where the high level optimizes the integer variables and the low level solves real-valued nonlinear programming problems [32]. The high level optimization problem can be formulated as

$$\begin{aligned} &\min_{x_{\text{in}}} f(x_{\text{in}}) \\ &\text{s.t. } \bar{h}(x_{\text{in}}) = 0 \text{ and } \bar{g}(x_{\text{in}}) \leq 0, \\ &x_{\text{in}} \in \mathbb{Z}^{n_{\text{in}}}, \end{aligned} \tag{B.21}$$

where $\bar{h}$ and $\bar{g}$ are the constraints which are only related with the integer variables, and the objection function value $f(x_{\text{in}})$ are calculated by the low level optimization. The optimization problem in the low level can be written as

$$\begin{aligned} f(x_{\text{in}}) = &\min_{x_{\text{re}}} f(x_{\text{in}}, x_{\text{re}}) \\ \text{s.t. } &\tilde{h}(x_{\text{in}}, x_{\text{re}}) = 0 \text{ and } \tilde{g}(x_{\text{in}}, x_{\text{re}}) \leq 0, \\ &x_{\text{re}} \in \mathbb{R}^{n_{\text{re}}}, \end{aligned} \tag{B.22}$$

where $\tilde{h}$ and $\tilde{g}$ are the constraints which are related with the real-valued variables and the integer variables. A branch-and-bound method, exhaustive method, etc. can be applied to the high-level integer programming, while a (multi-start) SQP, pattern search, etc. can be employed by the low-level nonlinear programming.

The MINLP problems can be solved by several existing commercial and free solvers [29], such as MINLP BB, SCIP, NOMAD, and BONMIN.

References

1. Becerra V (2010) Solving complex optimal control problems at no cost with psopt. In: Proceedings of IEEE multi-conference on systems and control, Yokohama, Japan, pp 1391–1396
2. Ross I, Fahroo F (2004) Pseudospectral knotting methods for solving optimal control problems. J Guidance, Control, Dyn 27:397–405
3. Gong Q, Ross I, Kang W, Fahroo F (2008) Connections between the covector mapping theorem and convergence of pseudospectral methods for optimal control. Comput Optim Appl 41:307–335
4. Canuto C, Hussaini M, Quarteroni A, Zang T (1988) Spectral methods in fluid dynamics. Springer, New York
5. Elnagar G, Kazemi M, Razzaghi M (1995) The pseudospectral Legendre method for discretizing optimal control problems. IEEE Trans Autom Control 40:1793–1796
6. Gong Q, Kang W, Bedrossian N, Fahroo F, Sekhavat P, Bollino K (2007) Pseudospectral optimal control for military and industrial applications. In: Proceedings of the 46th IEEE conference on decision and control, LA, USA, New Orleans, pp 4128–4142
7. Rutquist P, Edvall M (2008) PROPT: MATLAB optimal control software. Tomlab Optimization Inc., Pullman
8. Betts J (2002) A direct approach to solving optimal control problems. Comput Sci Eng 1:73–75
9. Stryk O (2000) User's guide for DIRCOL (version 2.1): a direct collocation method for the numerical solution of optimal control problems. Technical Report. Technical University of Munich. Munich, Germany
10. Ross I (2004) User's manual for DIDO: a MATLAB application package for solving optimal control problems. Tomlab Optimization Inc., Pullman
11. Rao A, Benson D, Darby C, Patterson M, Francolin C, Sanders I (2010) Algorithm 902: Gpops, a matlab software for solving multiple-phase optimal control problems using the Gauss pseudospectral method. ACM Trans Math Softw 37:22:1–22:39
12. Gill P, Murray W, Saunders M (2002) SNOPT: an SQP algorithm for large-scale constrained optimization. Soc Ind Appl Math J Optim 12:979–1006
13. Rao S (2009) Engineering optimization: theory and practice. Wiley, Hoboken
14. Boggs P, Tolle J (1995) Sequential quadratic programming. Acta Numerica 4:1–51
15. Nocedal J, Wright S (2006) Numerical optimization. Springer Science + Business Media, New York
16. Gill P, Murray W, Saunders M (2007) User's guide for SNOPT version 7: software for large scale nonlinear programming. Technical report. University of California, San Diego
17. Gill P, Murray W, Saunders M (1998) User's guide for NPSOL 5.0: a Fortran package for nonlinear programming. Technical report. University of California. San Diego
18. The Mathworks Inc. (1999) Optimization toolbox for use with Matlab: user's guide. The Mathworks Inc., Natick, MA, USA
19. Lewis R, Torczon V, Trosset M (2000) Direct search methods: then and now. J Comput Appl Math 124:191–207
20. The Mathworks Inc. (2004) Genetic algorithm and direct search toolbox for use with MATLAB: user's guide. The Mathworks Inc., Natick, MA, USA

21. Goldberg D (1989) Genetic algorithm in search, optimization and machine learning. Addison-Wesley, Reading
22. Vose M (1999) The simple genetic algorithm: foundations and theory. The MIT Press, Cambridge
23. Koehler L, Kraus W, Camponogara E (2011) Iterative quadratic optimiztion for the bus holding control problem. IEEE Trans Intell Transp Syst 12:1568–1575
24. Wang Y, De Schutter B, van den Boom T, Ning B, Tang T (2014) Efficient real-time train scheduling for urban rail transit systems using iterative convex programming. Technical report. Delft Center for Systems and Control, Delft University of Technology, Delft, The Netherlands. Submitted to a journal
25. Owen JH, Mehrotra S (2001) A disjunctive cutting plane procedure for general mixed-integer linear programs. Math Program 89:437–448
26. Dakin R (1965) A tree-search algorithm for mixed integer programming problems. Comput J 8:250–255
27. Linderoth J, Ralphs T (2005) Noncommercial software for mixed-integer linear programming. In: Karlof J (ed) Integer programming: theory and practice. Operations research series. CRC Press, Boca Raton, pp 253–303
28. Atamturk A, Savelsbergh M (2005) Integer-programming software systems. Ann Oper Res 140:67–124
29. Bussieck M, Vigerske S (2010) MINLP solver software. Wiley Encyclopedia of operations research and management science
30. Quesada I, Grossmann I (1992) An LP/NLP based branch and bound algorithm for convex MINLP optimization problems. Comput Chem Eng 16:937–947
31. Williams H (1999) Model building in mathematical programming, 4th edn. Wiley, Chichester
32. Wang Y, De Schutter B, van den Boom T, Ning B, Tang T (2014) Efficient bi-level approach for urban rail transit operation with stop-skipping. IEEE Trans Intell Transp Syst 15:2658–2670

Index

Y. Wang et al., *Optimal Trajectory Planning and Train Scheduling for Urban Rail Transit Systems*, Advances in Industrial Control,
DOI 10.1007/978-3-319-30889-0

MIX
Papier aus verantwortungsvollen Quellen
Paper from responsible sources
FSC® C105338

If you have any concerns about our products,
you can contact us on
ProductSafety@springernature.com

In case Publisher is established outside the EU,
the EU authorized representative is:
Springer Nature Customer Service Center GmbH
Europaplatz 3, 69115 Heidelberg, Germany

Printed by Libri Plureos GmbH
in Hamburg, Germany